Communication in Humans and Other Animals

Advances in Interaction Studies (AIS)

Advances in Interaction Studies (AIS) provides a forum for researchers to present excellent scholarly work in a variety of disciplines relevant to the advancement of knowledge in the field of interaction studies. The book series accompanies the journal *Interaction Studies: Social Behaviour and Communication in Biological and Artificial Systems*.

The book series allows the presentation of research in the forms of monographs or edited collections of peer-reviewed material in English.

For an overview of all books published in this series, please see
http://benjamins.com/catalog/ais

Editors

Kerstin Dautenhahn
The University of Hertfordshire

Angelo Cangelosi
University of Plymouth

Editorial Board

Henrik Christensen
Georgia Institute of Technology

Harold Gouzoules
Emory University

Takayuki Kanda
ATR, Kyoto

Tetsuro Matsuzawa
Kyoto University

Giorgio Metta
IIT, Genoa

Adam Miklosi
Eötvös Loránd University

Robert W. Mitchell
Eastern Kentucky University

Chrystopher L. Nehaniv
University of Hertfordshire

Stefano Nolfi
CNR, Rome

Pierre-Yves Oudeyer
INRIA, Bordeaux

Irene M. Pepperberg
Harvard University & Brandeis University

Kerstin Severinson Eklundh
KTH, Stockholm

Stefan Wermter
University of Hamburg

Volume 4

Communication in Humans and Other Animals
by Gisela Håkansson and Jennie Westander

Communication in Humans and Other Animals

Gisela Håkansson
Lund University

Jennie Westander
Parken Zoo/Linköping University

John Benjamins Publishing Company
Amsterdam / Philadelphia

The paper used in this publication meets the minimum requirements of
the American National Standard for Information Sciences – Permanence
of Paper for Printed Library Materials, ANSI z39.48-1984.

Library of Congress Cataloging-in-Publication Data

Håkansson, Gisela.
 Communication in humans and other animals / Gisela Håkansson, Jennie Westander.
 p. cm. (Advances in Interaction Studies, ISSN 1879-873X ; v. 4)
 Includes bibliographical references and index.
 1. Oral communication. 2. Visual communication. 3. Human-animal communication.
    4. Language acquisition. I. Title.
 P95.H35   2013
 302.2--dc23   2013006712
 ISBN 978 90 272 0457 8 (Hb ; alk. paper)
 ISBN 978 90 272 0458 5 (Pb ; alk. paper)
 ISBN 978 90 272 7201 0 (Eb)

© 2013 – John Benjamins B.V.
No part of this book may be reproduced in any form, by print, photoprint, microfilm, or any
other means, without written permission from the publisher.

John Benjamins Publishing Co. · P.O. Box 36224 · 1020 ME Amsterdam · The Netherlands
John Benjamins North America · P.O. Box 27519 · Philadelphia PA 19118-0519 · USA

Table of contents

Acknowledgements   XI

CHAPTER 1
Introduction   1
1.1 Introducing communicative behaviour   1
    1.1.1 Social life and the need of communication   2
    1.1.2 Linguistic perspectives on communication and language   4
    1.1.3 Biological perspectives on communication   6
    1.1.4 Comparing linguistic and biological perspectives   9
1.2 Major themes in comparisons between human and animal communication   10
    1.2.1 Hockett's design features   10
    1.2.2 Methodology and research questions   11
    1.2.3 Do animals use displacement?   12
    1.2.4 Do animals have combinatory principles?   14
    1.2.5 Do animal systems have arbitrary content/form relations? Do animals learn the system?   14
    1.2.6 How did language evolve?   18
1.3 Parent-offspring communication and cooperative breeding   19
1.4 Summary   22
1.5 Suggested readings   22

CHAPTER 2
Overview of biological signal systems   23
2.1 Introduction   23
2.2 Research methods and observational techniques   24
    2.2.1 Playback as a method for investigating communication   24
    2.2.2 Advances in technology influence the methods   25
2.3 Acoustic communication   26
    2.3.1 When and where?   27
    2.3.2 How? Production and perception   27
    2.3.3 Echolocation and other sounds in dolphins and whales   29

2.4 Visual communication 32
    2.4.1 When and where? 33
    2.4.2 How? Visual capacities 33
    2.4.3 Colour patterns 34
    2.4.4 Body postures and facial expressions 34
2.5 Tactile communication 35
    2.5.1 When and where? 35
    2.5.2 How? Skin, whiskers and sidelines 36
2.6 Indirect tactile signals – the seismic signal system 37
2.7 Chemical communication 38
    2.7.1 When and where? 38
    2.7.2 How? 39
    2.7.3 The olfactory system 41
2.8 Electrical communication 42
2.9 The multimodal honeybee – acoustic, visual, chemical, seismic communication 42
2.10 Summary 45
2.11 Suggested readings 45

CHAPTER 3
Human language – its use and learning   47
3.1 Introduction 47
    3.1.1 Social life and the languages of the world 48
    3.1.2 Observation techniques 51
    3.1.3 Language functions 53
3.2 The child's discovery of language – the first year of life 54
    3.2.1 Parent-infant interaction 54
    3.2.2 The "little universalist" – early perception of speech sounds 56
    3.2.3 Early vocalization – babbling 57
3.3 Language in the toddler 59
    3.3.1 Mapping forms to meanings 59
    3.3.2 Cultural differences reflected in children's language 60
    3.3.3 Words and world-views – what do you call your cousin? 62
    3.3.4 But what about grammar? 66
    3.3.5 Recursion 68
3.4 When problems arise – Specific Language Impairment 70
3.5 Second language acquisition – organizing language once again 72
3.6 Sign language – another modality 75
3.7 Summary 76
3.8 Suggested readings 77

CHAPTER 4
Human nonverbal communication   79
4.1 Introduction   79
     4.1.1 Research methods and observation techniques   80
4.2 Functions – what do we use nonverbal communication for?   81
     4.2.1 Permanent versus temporary expressions   82
     4.2.2 Synchronising in sympathy   83
4.3 Acoustic communication   84
     4.3.1 Extralinguistic expressions – laughters and cries   84
     4.3.2 Paralinguistic expressions – with speech   85
4.4 Visual communication   86
     4.4.1 Body postures   86
     4.4.2 Hand movements – pointing   86
     4.4.3 Other hand gestures   89
     4.4.4 Head and face   89
     4.4.5 Eyes and gaze   92
     4.4.6 Complexion   94
4.5 Tactile communication   94
4.6 Chemical communication   96
4.7 A model for analyzing gestures – The Ekman and Friesen model   98
4.8 Where verbal and nonverbal communication meet: Speech-related gestures   100
     4.8.1 The McNeill model   101
     4.8.2 Using knowledge of nonverbal expressions in verbal descriptions   103
4.9 When production is not automatized   106
4.10 Summary   107
4.11 Suggested readings   108

CHAPTER 5
Our closest relatives – nonhuman primates   109
5.1 Introduction   109
     5.1.1 The primate family   109
     5.1.2 Social life   110
     5.1.3 Studying nonhuman primates – how it all began   112
     5.1.4 Observation techniques   113
5.2 Communicative functions   115
     5.2.1 The referential function – first reported in vervet monkeys   115
     5.2.2 The social function – with focus on synchronization of behaviour   118

- 5.3 Acoustic communication   119
    - 5.3.1 Some vocalizations and their use   120
    - 5.3.2 Vocal learning in nonhuman primates   121
    - 5.3.3 Structural aspects of vocalizations – do nonhuman primates have syntax?   122
- 5.4 Visual communication   122
    - 5.4.1 Body postures   123
    - 5.4.2 Hand and arm gestures   123
    - 5.4.3 Face and gaze   125
- 5.5 Tactile communication – a lot of grooming   126
- 5.6 Chemical communication   127
- 5.7 Cultural/dialectal differences – results of social learning   128
- 5.8 Teaching human language to nonhuman primates   129
    - 5.8.1 Speech   129
    - 5.8.2 Sign language   130
    - 5.8.3 Plastic chips   132
    - 5.8.4 Computers and lexigrams   133
- 5.9 Discussion – primary versus secondary communication in nonhuman primates   135
    - 5.9.1 Primary communication – data base   135
    - 5.9.2 Primary communication – functions and structures   136
    - 5.9.3 Secondary communication – human language as a second variety   137
- 5.10 Summary   137
- 5.11 Suggested readings   138

CHAPTER 6

Man's best friend – the dog   139

- 6.1 Introduction   139
    - 6.1.1 The Canine class   140
    - 6.1.2 Social life of wolves and dogs   143
    - 6.1.3 Observation techniques   145
- 6.2 Functions of Canine communication   146
    - 6.2.1 Social functions   146
    - 6.2.2 Studies of the referential function   147
- 6.3 Acoustic communication   148
    - 6.3.1 Barking   150
    - 6.3.2 Growling   151
    - 6.3.3 Chorus howling in wolves   151

6.4 Visual communication 152
    6.4.1 Head and face 152
    6.4.2 Tail 153
6.5 Tactile communication 154
6.6 Chemical communication 154
6.7 Developmental patterns in intra-specific communication 155
6.8 The dog in the human family – learning to communicate with another species 156
    6.8.1 Dogs' understanding of humans 157
    6.8.2 Humans' understanding of dogs 158
    6.8.3 Not interactional synchrony – but accommodation 158
6.9 Summary 160
6.10 Suggested readings 160

CHAPTER 7
Communication in birds   163
7.1 Introduction 163
    7.1.1 The Aves class 163
    7.1.2 Social life of birds 165
    7.1.3 Observation techniques 166
7.2 Functions – why do birds communicate? 167
    7.2.1 Alignment of songs – counter-singing and duetting 168
    7.2.2 Referential function in birds – the calls of the domestic fowl 169
7.3 Acoustic communication 171
    7.3.1 Variation across and within species 172
    7.3.2 Structural aspects of song 174
7.4 Learning how to sing 176
    7.4.1 Sensitive phases and developmental stages 177
    7.4.2 Dialectal differences 181
7.5 Visual communication 182
    7.5.1 Talking with the tail 182
7.6 Chemical communication 183
7.7 Birds and humans 183
    7.7.1 Teaching language to birds 183
7.8 Summary 186
7.9 Suggested readings 186

CHAPTER 8
Discussion and outlook – why language? 187
8.1 Introduction 187
 8.1.1 Social life and communication in humans and animals 188
8.2 What is so special about language? Revisiting Hockett's predictions 189
8.3 Why and how did language evolve? 191
 8.3.1 Why and when did it happen? 191
 8.3.2 Are there any parallel changes in human physiology? 193
 8.3.3 Is language a result of cooperative breeding? 195
8.4 Learning intraspecific communication – not only for humans 195
 8.4.1 Child language and animal communication 196
8.5 Can language be taught to nonhumans? 198
8.6 Summary 199
8.7 Suggested readings 199

Glossary 201
References 205
Index 241

Acknowledgements

This book is the result of a long journey, with many turnabouts and new directions. However, we have not travelled alone. Over the years, many friends and colleagues have given their support and constructive criticism on individual chapters or the entire manuscript. We would like to thank Elainie Alenkaer Madsen, Dorothy Bishop, Christer Blomqvist, Sara Farshchi, Marianne Gullberg, Dennis Hasselquist, Arthur Holmer, Per Jensen, Sverker Johansson, Per Lindblad, Bengt Sigurd and Jan-Olof Svantesson. Some chapters have been used as course texts and we thank all students who have provided useful feedback. Any remaining errors or omissions are our responsibility.

Finally, we also want to thank the series editor Angelo Cangelosi for encouragement and feedback.

Gisela Håkansson and Jennie Westander

Gisela Håkansson	PhD, professor in general linguistics at Lund University, Sweden
Jennie Westander	PhD, adjunct lecturer in biology at Linköping University and head of education and research at Parken Zoo, Sweden

Chapter 1

Introduction

1.1 Introducing communicative behaviour

In this book a linguist and a biologist join forces to bring together studies of communicative behaviour in different species. We present research on communicative functions, communcative forms and communicative development. Our aim is to bridge the gap between linguistics and biology and discuss communication within one coherent framework. This gives a unique perspective on communication including a broad spectrum of species. There is a large body of studies of social behaviour and communication, but they have not been compiled and presented within the same framework before.

An important point of departure is the shift of paradigm from the signaller-message-receiver perspective to a dynamic and interactive view of communication, expressed both in linguistics (e.g. Pickering & Garrod 2004, 2006, Linell 2009) and in biology (e.g. McGregor & Peake 2000, McGregor 2005). This shift opens up for a new way to compare human and animal communication. Comparisons will be on more equal terms when the focus of interest is not on how the message is created "in the head", but on how meanings are created in the context of interaction – since we can observe the interaction, but don't know what is "in the head" of other species.

In this initial chapter we introduce some of the central issues generally discussed in studies of communication in humans and animals. Such comparisons often have the human language as starting point and compare communicative systems of other species to verbal language of humans. However, this does not necessarily imply that we see human language as the "best" way of communication, but simply that it is the system that we know best.

Our aim is to describe communicative systems as they are, but we will also comment on evolutionary aspects. Needless to say, the purpose is not to give a complete picture of the communication of humans and other animals. That would take more than could be captured within a single volume. What we hope is to give glimpses showing the richness in the communicative systems of different species – and to inspire the reader to read more. For brevity, when comparing different systems we will use the terms "humans" and "animals" as a shorthand notation system in most chapters, even though we are aware of the fact that also humans are animals.

The chapters are organized as follows. In this introductory chapter, we give a general idea of the main topics and how research in linguistics and biology can be combined to form a unified view of the area. Chapter 2 gives a short outline of general biological signal systems, with the respective sensory channels as points of departure. Chapter 3 deals with human verbal language and language acquisition. Chapter 4 treats human nonverbal communication. After that, there are three chapters presenting communication in some selected species: nonhuman primates (Chapter 5), dogs (Chapter 6), and birds (Chapter 7). We have chosen species where there are many studies on communicative behaviour, which results in both homologous comparisons (humans and nonhuman primates share a common origin) and analogous comparisons (dogs and birds share some behaviours with humans). All chapters follow the same general outline, starting with communicative functions, then communicative channels and forms of expression. Whenever possible – that is when there is data on development – our focus will be on how individuals proceed from not being able to communicate to reaching the state of competent communicators. The final chapter, Chapter 8, provides a synthesis of communication in humans and other animals, where findings from different perspectives are integrated and evolutionary aspects are taken into consideration. Since the book has a cross-disciplinary nature, we have added a Glossary section at the very end. This will help students of different disciplines to understand basic concepts in biology and linguistics.

But first, let us begin by defining the concept *communication*. The etymology of the term is Latin *communicare*, which means 'sharing' or 'transferring'. There are many different views and opinions about what it means to *communicate* – from the suggestion that it has to do with intentional sending off a message to a receiver (like writing and sending off a letter), to the idea that it has more to do with some kind of interactional synchrony where the participants cooperate and align to each other to create meanings and reach an understanding. Our interpretation is broad and closest to the last one: we define communication as interaction between individuals, taking place in dyads and in networks. Communication is a complex process embedded in a particular context, and the meanings are often parts of the context. In the following sections we will first give some examples of communicative behaviour in different social groups, and then a brief historical overview of how the topic has been approached by linguists and biologists over the years.

1.1.1 Social life and the need of communication

Why do we communicate? What do we need to share with others? Most animals have some kind of social interaction with another member of the same species, if only in order to find a partner to breed with, but there are huge differences when it comes to type and amount of communicative behaviours. Group life is the obvious setting for rich communication. Human groups are characterized by a constant flow of verbal and nonverbal communication; people are talking, listening, smiling, gesturing, touching, and laughing. Many other primate groups, such as chimpanzees (*Pan troglodyte*) and bonobos (*Pan paniscus*) also display constant social interaction – they vocalize, play, hug and groom each other. The social behaviour of wolves (*Canis lupus*) and dogs (*Canis familiaris*) exhibit the same continuous communication of feelings and states, in their case by vocalizations, body postures, facial expressions, sniffing and licking. Primates (including humans, of course) and Canids (e.g. wolves and dogs) are examples of individualized groups, where members recognize and know each other, have complex greeting ceremonies and other formalised means of communicating group membership.

At the other end of the continuum, there are animals that lead solitary lives, such as the tiger (*Panthera tigris*). The social life of the tiger is mostly about bringing up the offspring. The adult female and male meet to mate, but spend the main part of their lives apart. The cubs stay with their mother for around two years and then they are independent enough to be on their own. Tigers do keep track of others, however, by the scent markers they leave on territory borders (Smith et al. 1989).

There are also examples of seasonal social life. Some species vary their social life according to the seasons, for example the walrus that get together in aggregations of thousands of individuals during late summer, but in smaller groups during winter.

The social system that has been discussed the most in studies on communication is the *fission-fusion* social organization. This has been suggested to be the system behind the early development of human cultures. The term fission-fusion was originally suggested by Kummer (1971) to describe the social organization of hamadryas baboon (*Papio hamadryas*) groups. Kummer identified groups at different levels. First of all there are small *family* units that get together into larger *clans*, then the clans may joined together and constitute *bands*, and the bands can unite and form *troops* of several hundred individuals who spend the night at the same cliff. Not everyone can leave one group and join another, but the social hierarchy determines who goes with whom. Typical in a fission-fusion system is that individuals of a larger group regularly split into subgroups and then merge together again. This generates a need to communicate about when and where to travel and whom to follow, as well as how to re-establish relationships when the group meet again. Splits and reunions of this type happen in the daily life of chimpanzees (Boesch & Boesch-Achermann 2000), humans (Rodseth et al. 1991), some birds (e.g. Coropassi & Bradbury 2006), dolphins (Lusseau et al. 2006) and whales (Rendell & Whitehead 2001). Individuals in a fission-fusion system have to develop safe means for recognition, showing that they belong together. Many species use vocal convergence to accomplish this goal – they develop some kind of group dialects.

There are both positive and negative sides of living in groups. On the one hand it gives protection, on the other hand it is harder to hide for predators. Establishing and keeping a rank order may cause aggressive encounters, and in times when there is not enough to eat the group may have to disperse. In order to keep a group together and avoid aggression, many species have developed complex communicative behaviour. Basic communicative needs are for example to find a partner to mate with, take care of the offspring, warn against enemies, announce when and where to travel, and where to go to find food. In order to meet these needs behaviours with highly specialized functions are developed, such as contact calls and alarm calls, signals for dominance, submission, and reconciliation, ways to announce the reproduction status and good food sources. Besides these signals, there is a synchronization of behaviours in many mammals for feeding and vigilance.

1.1.2 Linguistic perspectives on communication and language

The word linguistic is derived from latin *lingua*, 'tongue', and refers to the study of human language. Linguistic research on communication generally concerns the use of words and sentences in different human groups. Human nonverbal communication, for example gestures and body postures, are rarely dealt with in linguistics departments, but left to psychologists. The area of linguistics has undergone dramatic changes over the centuries – from focussing on grammar and meaning as objects in their own right, to studying language production and perception, acquisition and use in communicative interaction.

The first studies on language are more than two thousand years old. The grammatical descriptions of Sanskrit (Panini 500 BC) and Greek (Aristotele 300 BC) laid the foundation to how we see grammar even today. Panini's ideas on phonology and Aristotele's suggestion to divide sentences into two parts, predicate and subject, are still valid as basis of linguistic descriptions. From the Greek philosophers came also a strong interest in *meaning*. Is there a relation between the meaning of a word and its form? Are words natural reflections of reality? (This philosophical way of looking at meaning is not used within linguistics today, but instead we see meaning as being created in the specific context, see Chapter 3).

Linguistic thinking took a new turn in the 19th century, following the spirit of the time, and under the influence of Darwin's evolutionary perspective. The aim was to study language change and to try and reconstruct a proto-language on the basis of differences across modern languages. For example, the finding of systematic differences in the first consonant [p] in Latin *piscis, pater* and *pes* and the [f] in Germanic *fish, father* and *foot* (Grimm's law), demonstrates that a sound change has taken place over time, differentiating the Romance and German language families. Such changes suggest that there might have been a common Proto Indo-European language (PIE).

This way of classifying languages on structural grounds on languages parallels the biological view on evolution of species.

A new perspective on linguistics opened up with the works of the Swiss linguist Ferdinand de Saussure (1857–1913). He has been suggested to be the father of linguistics, and he revolutionized linguistics by moving from the historical aspects to freezing language in time. He put forward three suggestions, (1) that a language can be described independently of its historical origin, (2) that there is a difference between *la langue* (the underlying system) and *la parole* (the actual utterances), and (3) that there is an arbitrary relationship between the concept and the form of a word. He defined language as a rule-governed symbolic system, with arbitrary relationships between content and form. The concept of 🐴 can have the form 'cheval' (French), 'horse' (English), 'hevonen' (Finnish), 'Pferd' (German), 'hippos' (Greek), 'mori' (Mongolian) etc. No word form is more natural than another. This answers one of the questions from the Antique philosophers (about the naturalness of the form) and gives way to the idea that different languages can have different ways to divide up the world (see Chapter 3). From now on it became common to freeze language and analyse it without always trying to find historical commonalities. A distinction between the inner system and the actual use of language is also taken up by the America linguist Noam Chomsky, who suggests that the optimal way to study language is to isolate and examine the linguistic *competence*. i.e. the internal representation of grammar, apart from the *performance*, its actual use. Chomsky also revolutionized linguistics by formalizing grammar to describe not only the grammar of a particular language but also underlying grammatical relations and categories. His *Syntactic Structures* (1957) focuses on grammatical structures that constant across different languages, a Universal Grammar. This perspective emphasizes the cognitive side of language.

Research on how language is used during the process of communication was long inspired by the work of Shannon and Weaver (1949) on information transmission, used in telecommunication models. The idea is that the transmission involves a message, a sender and a receiver. A message is supposed to be translated from thought into language and then sent to a receiver. This would imply an intention from the side of the sender to send a particular message to a receiver. Today, many researchers have a more complex view on what happens during a communicative situation, and suggest that the different participants have a share in the communicative event. For example, according to the interactive-alignment model of speech production (Pickering & Garrod 2006) speakers tend to reuse words and constructions they have just heard from others in order to reach a successful communication (similarly to the synchronization behaviour found in nonverbal communication, see Chapter 4).

Linguistic studies of language functions were long in the background. By making a distinction between *la langue* and *la parole* Saussure instigated the study of variation in time, how language varies among its users, but analyses of how language is actually used in interaction did not come about until Roman Jakobson's work in the 1960's. Jakobson suggested different language functions focussing on the message (the informative function), the speaker (the expressive function) and the listener (the social function), respectively. Talking about different functions of language shows that not all communication has to do with transfer of information, but there may be other goals. But it is still mostly about the speaker, the speaker sends information, the speaker expresses emotions and the speaker gains socially. This focus on communication from the perspective of the speaker has been criticized for giving a too simplified and one-sided view of communication (e.g. Linell 2009). From an interactive point of view, communication is at least two-sided; if there are no listeners there is no communication. Senders may send information that nobody picks up, and senders can give out information without intending to share it – since the receivers can read meanings that the sender does not even know is being communicated! This focus on the communicative event has repercussions on how *meaning* is interpreted. Instead of seeing word meanings as absolute, meaning is understood as something that is created in the interaction, a results of negotiation. The study of language as cooperative interaction involves researchers from other fields than linguistics, for example anthropologists, biologists, psychologists and sociologists.

As we will see in the following sections, however, the sender-receiver metaphor for communication has had a profound effect on comparisons between human and nonhuman communication, and it has often led to a focus on human-animal differences in terms of goal-directed intentionality.

1.1.3 Biological perspectives on communication

The word biology comes from Greek *bios* 'life' and *logos* 'the study of'. Biology deals with the study of life and living organisms, their structure, function and development, including behaviour such as communication. The British naturalist John Ray is known as the pioneer in the tradition of field biology with his study of instinctive behaviour in birds (1691). But it was another British naturalist, Charles Darwin (1809–1882), who transformed the whole study of animal behaviour. He undertook a five-year long trip around the world to explore and document the flora and fauna in different parts of the world. The diversity that he found led him to assume that the different species had descended from the same ancestors, and he suggested that the changes were intermediate solutions due to natural selection. In his book *The Expression of the Emotions in Man and Animals* (1872/1965) Darwin discusses communicative expressions in human and animals along the same lines – having a common origin – as he could observe many similarities in nonverbal expressions. Moreover, he sees no reason why human language would not obey the principles of evolution, but he proposes that language should also have developed by intermediate steps. Darwin's enormous influence on the study of animal communication and social behaviour is apparent in the interest in evolutionary explanations, not only in the 19th century but also today. Communication is rarely described only as behaviour by its own right, but also with regard to the selective advantage it might have had. Both phylogeny (the development of the species) and ontogeny (the development of the individual) are important. Studying the developmental process of the individual makes it possible to discern which behaviours are innate and which are learned, and which key stimuli release which behaviours. Ethologists study the biology of behaviour, and as in other biological studies of animals, one aim is to identify which characteristics are typical to a particular species. Just like physiological features can be listed, behaviours can be listed in so-called *ethograms*, where the behavioural repertoire of the species is described.

In the 20th century, two different schools of animal behaviour study developed. One is the North American behaviourist school, working with controlled experiments in laboratory settings, aiming for finding general learning principles. One example is the famous reinforcement experiment by Skinner (1938). The experiment demonstrated that a rat could learn a new behaviour by positive reinforcement. The rat soon learned to press a handle to obtain food, and then it pressed the handle more and more often. The other school of animal behaviour study, the European school, is based on observations in natural environments, often focussing on adaptation to the environment. One example is the classical study by the British zoologist William Thorpe (cf. Chapter 7) where he discovered dialects in the chaffinch (*Fringilla coelebs*; Thorpe 1958).

Ethology (from Greek *ethos*, 'character' and *logos*, 'the study of') as discipline became known to the larger public in 1973, when Karl von Frisch, Konrad Lorenz and Nikolaas Tinbergen received the Nobel Prize in Physiology and Medicine for their work on animal behaviour. These three scholars are often regarded as the founding fathers of ethology. The Austrian Karl von Frisch (1886–1983) specialized in the waggle dance of the bees (e.g. von Frisch 1954), Konrad Lorenz (1903–1989), also from Austria, discussed instinctive behaviour, fixed action patterns and imprinting (e.g. Lorenz 1981), and the Dutch ornithologist Nikolaas Tinbergen (1907–1988) experimented with key stimuli in herring gulls (*Larus argentatus*) in natural settings (e.g. Tinbergen 1951). This branch of biology was so new and unexplored that the Nobel Prize seems to have come as a surprise, at least for Tinbergen. He said in his Nobel Lecture (1973) that "Many of us have been surprised at the unconventional decision of the Nobel Foundation to award this year's prize for 'Physiology or Medicine' to three men who had until recently been regarded as 'mere animal watchers.'"

Tinbergen suggested four questions for the study of animal behaviour:

Box 1.1 Tinbergen's four questions (Tinbergen 1951)

1. What is the physiological causation, or mechanism, of the behaviour (for example, which is the releasing stimulus, or which hormones are influencing)?
2. What are the functions and the survival value of the behaviour? (adaptation)
3. How does the behaviour develop in the individual (ontogeny)?
4. How did the behaviour evolve in the species (phylogeny)?

Tinbergen himself spent much of his research on the first question, the mechanism, for example in the examination of innate behaviours and nervous system of herring gulls. The four questions are still relevant in the study of animal behaviour and communication. They are not restricted to ethological research, but stretch out to other disciplines as well. There is a large bulk of research on the questions of development and evolution within evolutionary biology, molecular biology and behavioural ecology as well as ethology. The behavioural ecologists bring general models of cost-benefit analyses and efficiency of behaviour into the descriptions. Krebs and Dawkins (1984) argue that there is a difference between signals that are mutually beneficial to the animals involved and signals that are used for exploitation. Those that are cooperative have a better balance between economy and detectability, whereas signals that are manipulative are conspicuous and repetitive (1984: 401).

Ethology also includes a sub-section called applied ethology. Here the knowledge about animal behaviour comes to practical use in our handling of animals kept in farms and zoos. In applied ethology, all Tinbergen's questions are studied: the causation and ontogeny are important in order to understand how stereotypical behaviour emerge; functions and phylogeny of behaviours help us to identify processes in domestication.

Another perspective on animal communication is provided in the network theory (McGregor & Peake 2000, McGregor 2005) where a dynamic picture of communication is given, focussing on the social interaction and taking new phenomena into account such as audience design (how expressions are modified to the audience) and eavesdropping (when individuals pick up information that was not intended for them). The communicative network paradigm is characterised by:

- many participants
- dynamic interaction pattern
  - the indirect addressee (the third party) can become the direct addressee
  - the direct addressee (the second party) can become the indirect addressee
- individuals in regular contact
- eavesdropping
- audience design

1.1.4 Comparing linguistic and biological perspectives

It is clear that the views on communication differ between linguists and biologists, as well as between schools within linguistics and biology. Both disciplines take interest in forms, functions and development of communication but in very different ways. It reminds us of the Hindoo fable of the blind men and the elephant, each man judging the animal from his own standpoint, as illustrated in the poem by Saxe (1872) in Box 1.2.

Box 1.2 John Godfrey Saxe: The blind men and the elephant

THE BLIND MEN AND THE ELEPHANT. A HINDOO FABLE

It was six men of Indostan
To learning much inclined,
Who went to see the Elephant
(Though all of them were blind),
That each by observation
Might satisfy his mind.

The First approached the Elephant,
And happening to fall
Against his broad and sturdy side,
At once began to bawl:
"God bless me! – but the Elephant
Is very like a wall!"

The Second, feeling of the tusk
Cried, "Ho! What have we here
So very round and smooth and sharp?
To me 'tis mightly clear
This wonder of an Elephant
Is very like a spear!"

The Third approached the animal
And happening to take
The squirming trunk within his hands,
Thus boldly up and spake:
"I see," quoth he, "The Elephant
Is very like a snake!"

The Fourth reached out an eager hand,
And felt about the knee
"What most this wondrous beast is like
I mightly plain," quoth he;
"'Tis clear enough the Elephant
Is very like a tree!"

The Fifth, who chanced to touch the ear,
Said: "E'en the blindest man
Can tell what this resembles most;
Deny the fact who can
This marvel of an Elephant
Is very like a fan!"

The Sixth no sooner had begun
About the beast to grope
Than, seizing on the swinging tail
That fell within his scope,
"I see," quoth he, "the Elephant
Is very like a rope!"

And so these men of Indostan
Disputed loud and long
Each in hos own opinion
Exceeding stiff and strong
Though each was partly in the right
And all were in the wrong!

Moral
So oft in theologic wars
The disputants, I ween
Rail on in utter ignorance
Of what each other mean
And prate about an Elephant
Not one of them has seen!

The question in the fable above is on what characterizes an elephant, and depending on perspective, markedly different definitions are given. In our discussion the question is about what constitutes language and communication, and the definitions from varying perspectives seem to be as disparate as in the fable. Language/communication is innate *and* learned, intentional *and* un-intentional, informative *and* expressive, can have fixed forms *and* be created in the dynamics of the context.

In our endeavour to compile as many issues, research question and results from linguistics and biology as possible to give a coherent pictures of communication, there are some lines of research that will be given prominence over others. For example, we will describe how communication functions and develops in the individuals and not go into depth with the evolution of communication in the species. However, we will return to such issues in the final chapter.

1.2 Major themes in comparisons between human and animal communication

Despite (or maybe encouraged by) all the problems with diverging perspectives and research paradigms, there is a large interest in comparing communication in humans and animals, and there has been an outburst of studies during the last decades. New questions and methods have emerged, and many of the old assumptions about human – nonhuman differences have been challenged. We will disentangle some of the properties suggested to be unique to humans and see if there are parallels also in nonhuman communication.

1.2.1 Hockett's design features

The most quoted definition of human language is the one suggested by the American linguist Charles Hockett (1968). He proposes a set of sixteen design features for the definition of language. Only if all these are present simultaneously, the communicative system may be called a language. The features are (1) using the vocal-auditory channel; (2) broadcast transmission with directional reception; (3) rapid fading; (4) interchangeable; (5) complete feedback; (6) specialization; (7) semanticity; (8) arbitrariness; (9) discreteness; (10) displacement; (11) openness; (12) tradition; (13) duality of patterning; (14) prevarication; (15) reflexiveness; and (16) learnability (Hockett 1968: 8–13). These features describe some typical features of human language, for example that it is vocal, directional, rapidly fading, speakers can change roles and they can hear what they are saying, it expresses meanings through combinations of units, it is passed on to others by tradition, etc.

Some of Hockett's design features turn up more often than others in the discussion of communication among humans and animals. We have chosen to discuss four of these: *displacement, duality of patterning, arbitrariness* and *learning*.

1. *Displacement* refers to the potential of language to go beyond the here-and-now. Human language allows us to talk about phenomena outside the immediate context, and to refer to things remote in time or space. We can also use language to create fiction and tell lies.
2. *Duality of patterning* refers to the fact that languages have units that can be combined and re-used in various ways. By replacing the (by itself meaningless) sound [p] in pig by [b], [d] or [w], we get new meaningful words like *big, dig, wig*. By adding the plural ending –s to *book*, we get a new meaning of the word *books*.
3. The *arbitrary* relationship between content and form means that a particular content can have different forms in different languages. As mentioned above, no form is more natural or logical than the other (even if a native speaker of English may perceive 'horse' as the most natural way to talk about a *Equus caballus*).
4. The arbitrary relationship has implications for the acquisition of language – forms have to be *learned* for each language.

Each of these features are found in nonhuman communication, but rarely all at the same time, and often only to a small extent.

1.2.2 Methodology and research questions

Before we proceed to describing studies on communication in humans and animals, we will say a few words about methodology. The comparison between humans and nonhumans is not as easy as it might seem at first sight, and there are many methodological pitfalls (see Rodseth et al. 1991). For example, which species should be selected to represent "nonhumans" in the comparison? Bees, birds, crocodiles, or nonhuman primates? If the aim is to find similarities, the obvious choice is our closest relatives, the chimpanzee or the bonobo. If the aim were to stress differences, then maybe crocodiles would be a better choice. However, even if we agree to use our nearest relatives in the comparison it is not unproblematic which group to choose. In the early days of describing animal behaviours, a single individual could be thought of as representing the whole species (e.g. the study on one chimpanzee by Ladygina-Kohts 1935/2002). However, today we are aware of both individual and group differences, for example in chimpanzee mother-offspring communication (Bard et al. 2005). This means that we cannot base a comparison on only one population of chimpanzees.

Furthermore, variation is not only measured in a qualitative way – different behaviours in different groups – but it can also be looked upon quantitatively, i.e. differences in frequency of use. As pointed out by Thorpe (1972), it is often a continuum of frequency of use. If a feature is found to occur in two species, it does not follow that it is used in the same way. For example, in human language words have specific meanings – and many other animals have expressions with specified meanings, such as warning calls against certain predators, food calls specifying which kind of food. The size of vocabulary in animals, however, does not match that of a human two-year old, who has a lexicon of more than a hundred words, at least not to what we know so far.

The obvious problems in finding good candidates to compare across species have not stopped researchers from comparing, but there is an increasing amount of comparative studies from a large number of aspects. In the following, we will present in more detail some of the "hot topics" over the years, such as: Can animals use displacement and communicate about things that are remote in time and space? Do animal communicative systems have combinatory elements? Do animal communictive systems have arbitrary relations between form and meaning, which have to be learned? How did language evolve?

1.2.3 Do animals use displacement?

In relation to the question about displacement, in the sense of going outside of time and space, it is necessary to introduce the concept *symbolic* meanings in language. There is a long-standing discussion about what symbols really mean and whether they are ever used by nonhumans. We will try to make this story short. The very instantiation of a symbolic system is the human language. It is based on the use of *words*, hence the term verbal communication: *verbum* is Latin for *word*. Words carry *symbolic* meaning, in the sense that we may *refer* to external phenomena. We use words to talk about events, objects and persons. We can also use language for displacement – we can talk about events and people remote in time or space. Thus, language allows us to go outside the here-and-now. It is even possible to talk about events and objects that do not exist. The Russian linguist Roman Jakobson defined the *referential function* of language as something related to truth values (e.g. Jakobson 1960), but we can not only talk about things that are true but also about phenomena that we know belong to a fantasy world.

The mental pictures of the persons, objects or events talked about are called *representations*. Representations can be more and less abstract. Some are directly cued by something in the environment, whereas others are detached from the immediate context, that is not triggered in the situation but belonging to an inner world (Gärdenfors 1996, 2004). When discussing representations, we move away from the area of communication into the area of cognition. Another concept that is related to representations is object permanence: the ability to keep an inner image of an object when the object has disappeared out of sight (Piaget 1959). We will not go into detail about the many definitions, but simply give a list of some labels commonly used for communication about external phenomena: for example *referential function* (Jakobson 1960), *representational function* (Halliday 1975), *symbolic reference* (Deacon 1997), *environmental signals* (Bradbury & Vehrenkamp 1998), or *semantic signals* (Altmann 1967, Seyfarth et al. 1980a, b). In this book we will use *referential communication* (for a semiotic view on the use of the terms symbols, signs and representations, see Sonesson 2006).

The debate about whether other species than humans have access to representations has preoccupied researchers on communication in humans as well as in other animals over the years (for overviews see Gallistel 1990, Marler et al. 1992). For a long time, referential communication was seen as absolutely impossible in animals. However, this view was challenged by the discovery of specific alarm calls in vervet monkeys (*Cercopithecus aethiops johnstoni*) in Kenya (Struhsaker 1967). The calls were identified as referential, since they pointed to external referents and were not just expressions of inner emotions. By recording calls and play them back to the group (see the playback method, Chapter 2.2.1), it was possible for researchers to infer the meanings. It was found that the calls specified predator category by having different forms depending on predator; an eagle, a snake or a leopard. The 'eagle call' had the effect that the monkeys hid on the ground, whereas the 'snake call' made them run to the trees (Struhsaker 1967, Seyfarth et al. 1980a, b – see Chapter 5.2.1 for details). These and other similar findings inspired others to use the same research questions and methodology on other animals, and both alarm calls, referring to different predators (e.g Cheney & Seyfarth 1990, Gros-Louis 2004, Zuberbühler 2000a), and food calls, i.e. calls that inform the others that a food source have been identified (Gros-Lois 2004) were found. The referential function of alarm calls with predator specification have been found in vocalizations of many other species as well, for example Gunnison's prairie dog (*Cynomys gunnisoni;* Slobodchikoff et al. 1991), red squirrel (*Tamiasciurus hudsonicus*; Greene & Meagher 1998); yellow-bellied marmots (*Marmota flaviventris*; Blumstein & Armitage 1997), and redfronted lemur (*Eulemur fulvus rufus*; Fichtel 2004), domestic chicken (*Gallus gallus*; Marler et al. 1993, Forkman 2000) and Siberian jay (*Perisoreus infaustus*.; Griesser 2009). As we will see below, for many of these species, it has been possible to track down the developmental routes of the calls (Hollén & Radford 2009).

To summarize, the discovery that animal vocalizations have the potential to refer to external phenomena defies the assumption that referential meaning is what differentiates human from nonhuman communication. However, the picture is more complex. Even though the referential function is suggested as a prime human feature, humans do not use it all the time. There are also social and affective functions of language use, and, apart from informative events such as lectures, many contexts have a mixture of different functions (see Chapter 3.1.4). Thus, what we are seeing are multifunctional signals, where the same expression can carry both referential, social and affective meanings – in human as well as animal communication.

1.2.4 Do animals have combinatory principles?

Another important characteristic of the human language is its potential to combine units. The reason why we can have large vocabularies of tens of thousands words is because the words are constructed from discrete categories (or building blocks) which can be used and combined in a number of possible ways. For example, the word *rate* is distinguishable from *late* by the initial consonant; /r/ versus /l/. Each of these consonants lacks meaning of its own, but they get meaning when used in the combination. The principle that guides this has been labelled "double articulation" meaning that there is a structure in two dimensions, and that a small number of discrete units can form a large number of words. For example, the speech sounds of a language is typically around 30–50, and the vocabulary over 10 000 words. The combinatory principles are working on different levels. Not only can words be created from speech sounds, new words can also be created by combining old words (*sheep + dog = sheepdog; bull + dog = bulldog; gun + dog = gundog*). The possibility to create new words by combining existing words is a central property of language called *recursion* (also used in mathematics for functions that define their functions). In English, it applies to compound words, but more importantly to clause structure allowing for embedded clauses. For example, the clause *they got married* can be reformulated to *she said that they got married* or *he said that she said that they got married* or *I think that he said that she said* ... This can go on as long as we can keep the different persons and what they are saying in memory. Recursion has been suggested as the most important property of human language, the feature that distinguishes human language from all other communicative systems – the feature that makes human language unique (Hauser et al. 2002).

1.2.5 Do animal systems have arbitrary content/form relations? Do animals learn the system?

That human languages have arbitrary relationships between content and form is evident in the different ways to express the same things (eg. the word for horse is different in different languages). This implies that speakers must learn the specific forms (and contents) of their language, in order to communicate with others. But what about animals? Do they also learn their communicative signals?

Learnability is a much debated issue in human – animal comparisons. For a long time, it was believed that the main difference between human and animal communication was that humans learn their languages, while animals use innate expressions. So, for humans it was a question of nurture, and for animals it was a question of nature. This question was revitalized when Konrad Lorenz introduced the term *imprinting* (*Prägung* in German), in the classic article *Der Kumpan in der Umwelt des Vogels* (Lorenz 1935). Lorenz claimed that certain behaviours follow an innate perceptual sketch or outline, which is released by outer experience during a critical period in time. The result is *imprinting*. A famous example is the "following-reaction" in goslings. In nature, the first object that the gosling sees would be its mother, so the imprinting would have a practical value. However, by setting up experimental situations, Lorenz found that the gosling would follow almost any moving object, be it boxes, balls, other birds, or humans, and will stick to the first impression even after having met other options. This shows that outer experience can release the behaviour, and that the behaviour is imprinted. Lorenz (1935) gives two qualities that characterize true imprinting:

Box 1.3 Konrad Lorenz on imprinting

> **IMPRINTING**
>
> 1. The process is confined to a very definite period of individual life, a period which in many cases is of extremely short duration; the period during which the young partridge gets its reactions of following the parent birds conditioned to their object, lasts literally but a few hours, beginning when the chick is drying off and ending before it is able to stand.
>
> 2. The process once accomplished, is totally irreversible, so that from then on, the reaction behaves exactly like an 'unconditioned' or purely instinctive response. This absolute rigidity is something we never find in behavior acquired by associative learning, which can be unlearned or changed, at least to a certain extent. (Lorenz 1935:264)

In the 1950's, two British zoologists, William Thorpe (1951, 1958) and Peter Marler (1952) presented groundbreaking results on innateness and sensitive periods in studies of the song learning of songbirds. The fact that some birds showed dialectal variation went against the earlier assumption that animal communication was innate. Technical innovations, such as the sound spectrograph, made it possible to describe the structure of the songs and to pinpoint exactly where the dialectal differences were. Experiments on chaffinches (*Fringilla coelebs*) raised in isolation showed that isolated birds only use a fraction of the song (the innate part), whereas birds exposed to adult input use the full song. Thorpe concludes that the "Full Chaffinch song is thus an integration of inborn and learned song patterns, the former constituting the basis for the latter" (Thorpe 1958:568). (For presentations of these and other experiments, see Chapter 7). In other words, the old idea of animal communication consisting of innate expressions, was challenged. What about the other idea, that human language is learned?

The findings from biology inspired Noam Chomsky, who saw a possible parallel to the inner program of birds, in inner programs in human children. Chomsky was at the time engaged in a debate with the behaviourists, arguing against their idea that language acquisition was a matter of imitation. Chomsky insisted that there was more to language acquisition than mere imitation.

Box 1.4 Noam Chomsky on "special design"

> **SPECIAL DESIGN FOR LANGUAGE ACQUISITION**
>
> The fact that all normal children acquire essentially comparable grammars of great complexity with remarkable rapidity suggests that human beings are somehow specially designed to do this, with data-handling or "hypothesis-formulating" ability of unknown character and complexity.
> (Chomsky 1957: 57)

Chomsky's idea of a "special design" was later to develop into a hypothesis of "Universal Grammar", assumed to be part of the nature of human children. This Universal Grammar (or UG) works like a constraint on possible grammars, which makes the task easier for children when they try to learn a language by bringing order in what people say.

Another linguist influenced by biology was Erik Lenneberg, whose book *Biological Foundations of Language* (1967) had a large impact on theories on language and language acquisition. Lenneberg suggested a critical period for language learning, starting around 2 years of age and ending at puberty. The evidence came from children with Down's syndrome, and from children that recover from traumatic aphasia (where "progress in language development was only recorded in children younger than 14." Lenneberg 1967: 155). Lenneberg attributed this to lateralization of language functions in the brain hemispheres, and that when language is localized in the left hemisphere after puberty, it cannot change.

Critical periods, or rather the less rigid term sensitive periods, have also been discussed in relation to second language learning and language evolution. The often experienced problems of learning additional languages, especially in school-settings, has led to the claim that the second language learner is outside the limits of a critical period, and therefore has no access to Universal Grammar (e.g. Bley-Vroman 1989). Hurford (1991) goes as far as claiming that second language learning is outside the perspective of language evolution, and that there was never a selective pressure to acquire more languages after puberty.

The idea of innateness and sensitive periods for children is appealing as explanation of the fact that all children acquire their first language in such a successful manner, whereas second language learners often fail to reach the highest levels. But how can it be empirically tested? For songbirds, the method has been to isolate young birds for certain periods, or to deafen them by surgery. For ethic reasons, this would not be possible with humans. Except for a few case studies with aphasia, and isolated children (e.g. *Genie*, Curtiss 1977), the empirical proof for a sensitive period in language acquisition is practically nonexistent.

Research on the earliest stages in language acquisition has revealed that babies seem to be born with so-called *categorical* perception, that is, they are able to discriminate between different speech sounds. This ability has been found to work already from the age of 4 days (Eimas et al. 1971). The sensitivity for speech sounds undergoes significant changes during the first year. Sounds that are not part of the surrounding language(s) get lost, and only the sounds that are used in the environment are distinguishable. This change takes place as children discover the form-meaning connection and acquire their first words (cf. Chapter 3.2).

The ability to discriminate speech sounds is not, however, exclusively for humans. Experimental studies have shown that not only human babies, but a number of animal species have the capacity for categorical perception. The first to show this were Kuhl and Miller (1975) who reported that chinchillas (*Chinchillas laniger*) were able to discriminate between the consonants /d/ and /t/. Subsequent studies showed that also nonhuman primates (Ramus et al. 2000), birds (Dooling et al. 1995), and dogs (Fukuzawa et al. 2005) have categorical perception. This suggests that the mechanism underlying categorical perception may be part of a general auditory system. It is not even unique to primates, since also other species have the capacity.

When animals show ability for categorical perception, they perform similarly to the first months in life for human infants – they are able to differentiate unknown speech sounds. They don't follow the next stage, when children tune in to the speech sounds of the surrounding language. Of course, since these animals are being tested on human speech sounds, all the sounds are unfamiliar. In the studies mentioned, the whole purpose was to see whether nonhumans are able to identify human phonemes and the experiments show that nonhuman primates, birds and dogs are able to discriminate /ra/ – /la/. We do not know whether these species use categorical perception in their own communicative system. To use sounds that are not part of our human system is difficult since we do not know the character of the sounds. A better way would be to identify which features are used in a discriminating way in these species and then set up experiments to test this. An example of such an experiment is the study by Weary (1989), where phrase length in the song of the great tit (*Parus major*) was tested. The results showed that length was not perceived categorically, but continuously. It may be that the type of length tested does not work in a categorical way, and that there are other features that should be tested. For suggestions of future research to solve this problem, see Yip (2006).

1.2.6 How did language evolve?

Not only the development in the individual, but also the evolution of communicative systems in the species is of interest (cf. Tinbergen's question 4). For reseachers on human language, the question of language evolution was for a long time a more or less forbidden area, and it was regarded as a meaningless issue. However, the interest in language evolution has seen a dramatic increase during the last decades. The paper *Natural language and natural selection* by Pinker and Bloom (1990) can be seen as the starting point, and after that there has been a steady flow of monographs and edited volumes covering the issue, e.g. Lieberman 1991, 2000, Deacon 1997, Hurford et al. 1998, King 1999, Lieberman 2000, Cangelosi & Parisi 2002, Jackendoff 2002, Wray 2002, Christiansen & Kirby 2003, Johansson 2005, Tallerman 2005, Tomasello 2008, Fitch 2010 (to mention some). Several new journals and conferences have emerged (e.g. in 1996 the international conference series on The Evolution of Human Language, Evolang, had its first meeting in Edinburgh).

When discussing the emergence of human language, there are two routes to chose between: continuity, i.e. that language evolved in intermediate steps from other communicative systems, or discontinuity, i.e. that language emerged suddenly.

The continuity hypothesis is represented by Darwin. Situating language within a social situation, where it is imperative to mate and breed, support a family, find alliances, defend territory, and share the life with others, it is easy to see that human basic communicative needs are not different from those of animals. Followers of this view are for example Dunbar (1996), who suggests that talking is the human equivalent to grooming, i.e. for social comfort, and Burling (2005), who proposes that language is to humans basically what the peacock's tail is to the peacock, namely the advertising of oneself.

The discontinuity proponents are widely represented in the linguistic literature. In the spirit of the 17th century French philosopher Descartes, the traditional view has been to postulate non-continuity between communication in human and other animals. The non-continuity supporters claim that language is uniquely for humans. Chomsky (e.g. 1957, 1965, 1980, 2000a, 2000b, 2004), the most influential linguist of our time, suggests that only humans have the capacity to pick up the intricate grammatical rules of a language in a natural environment since children are born with an innate schema to deal with grammars – a Universal Grammar. When Chomsky's ideas first came up, it was applauded as a confirmation of biology. Lorenz wrote: "A strong support for human ethology has come from the unexpected area of linguistics; Noam Chomsky and his school have demonstrated that the structure of logical thought – which is identical of syntactic language – is anchored in a genetic program" (Lorenz 1981:11). This idea of a genetic program is translated into a *language instinct* by Pinker (1994:18) who states that language is "a distinct piece of the biological makeup of our brains" and "no more a cultural invention than is upright posture". Another linguist arguing for discontinuity is Lenneberg (1967). He makes a case against a continuity theory and claims that there are neuropsychological prerequisites for language, and that certain language problems seems to be inherited. Empirical research on grammatical disorders has given some support to this observation (Wexler 2003 therefore calls the findings "Lenneberg's dream"). For example, family members from three generations have been found to exhibit the same grammatical problems (Gopnik 1990, Hurst et al. 1990), and a genetic mutation in the region FOXP2 has been found to play a role in language disorders (Lai et al. 2001). Interestingly enough, Scharff and Haesler (2005) point out that the same gene is important in the song learning of songbirds. (For more discussion on FOXP2, see Chapters 3 and 7).

Language being a highly complex phenomenon, it is not surprising to find that the proponents on continuity and proponents of non-continuity have different views on what constitutes language. The continuity proponents see language as a communicative act in a social context, whereas the discontinuity proponents focus on language as a vehicle for thinking. The continuity proponents focus on meaning and pragmatics, and the discontinuity proponents focus on grammar.

1.3 Parent-offspring communication and cooperative breeding

If communicative behaviours are not fully innate but also learned, how does this learning come about? Does it come from the early parent-offspring interaction? Direct evidence for the exact relation between parent-offspring interaction and learning outcome is sparse in other species than songbirds (and possibly humans), but a new interest during the last decennium has generated a rich literature on parent-offspring communication. In this section we will give some examples from the literature.

The initial state of the offspring differs between species and can be placed on a scale from altricial (born totally helpless) to precocial (born with some abilities). The demands on parent-offspring communication vary according to this. Songbirds with altricial offspring in a nest do not involve in communication with the young during the very first days, whereas domestic fowl for example, with precocious young who can walk by themselves, use vocalizations to help them find food (see Chapter 7).

A basic function of parent-offspring interaction is to learn how to recognize each other, that is *individual recognition*. This is important in, for example, seabirds that spend their infancy standing on a crowded cliff. Guillemots (*Uria aalge*), must learn the calls of their fathers and use this knowledge to find the father in the water, when they have taken the "big jump" from the breeding cliff down to the ocean (Tschanz 1959). There are also results describing the gradual process of learning. A study of the cliff swallow (*Hirundo pyrrhonota*) demonstrates how their vocal recognition develops over time. Cliff swallows do not recognize their parents' call at 9 days, but they are able to do so at 18 days of age (Beecher et al. 1985).

Reciprocal recognition occurs in some species but not in all. In herd-living mammals like ungulates, two different types of behaviour to avoid predators have evolved. Some ungulates are followers (e.g. sheep; *Ovis aries*) and others are hiders (e.g. fallow deer; *Dama dama*). The followers are mobile and can flee with the group whereas the hiders lie concealed in the vegetation waiting for the mother to come and feed them for the first 2–3 weeks in life. For sheep, vocal identification is reciprocal – the ewes recognize vocalizations from their own lambs, and lambs recognize vocalizations from the mothers (Sèbe et al. 2007). In the fallow deer however, the hiding fawn stays silent, so there is no need for mothers to learn to recognize the call. Experiments have shown that the fawn recognizes the mother's call, whereas the mother does not recognize her own offspring (Torriani et al. 2006).

In many animals, parent-offspring communication goes beyond mere individual recognition. In Nile crocodiles (*Crocodylus niloticus*), the young crocodiles communicate with parents already before hatching, inside the eggs. When they emit specific distress calls, the parents react in a way to help hatching and to bring them to the water. However, the calls are not individual, but they belong to a developmental scale. There is a gradual change in fundamental frequency in calls between hatching and four days of age, giving information about the age and size of the offspring. "A possible consequence is that a crocodile mother could be able to assess its offspring's age/size and adapt its behaviour accordingly" (Vergne et al. 2007: 53). Thus, calls from the infants influence parental behaviour.

Some parents are able to guide the behaviour of the offspring by vocalizations. The parents of young western sand pipers (*Calidris mauri*) do that in a sophisticated way. The chicks leave the nest and feed themselves after hatching, but they stay in the vicinity for 2–3 weeks. During this period the parents use different calls to make the chicks move away, to have them crouch on the ground, or to make them approach. Also the young chicks use calls – they give alarm calls when they are in danger, and they respond by contact call when the parents tell them to approach (Johnson et al. 2008).

Humans are socialized through a long period of so-called *cooperative breeding*. This means that the responsibility for the child is shared among members of the whole group. Since the responsibility is shared in the group, the infant must learn how to communicate with others than the biological parents, for example older siblings, grandparents, aunts and uncles. This implies that the children learn the different pronunciations of the language-specific speech sounds, produced by young and old, male and female voices. The American anthropologist Sarah Blaffer Hrdy (2009) suggests that it was the cooperative breeding that laid the foundation of human communication. Since human mothers do not carry the baby all the time, they use vocalizations as contact, when the child is put down. They can be said to "stay in touch without touch". Studies comparing different cultures have demonstrated that there is a relationship between body contact and language contact. In cultures where babies sleep by their mothers and have a lot of body contact, parents do not talk as much to their infants as in cultures where infants sleep by themselves. In the early parent-infant interaction both auditory and visual elements have been described, for example that adults use a particular way of speaking (the "Baby Talk register"; cf. 3.2) and that there is a synchronization in the interaction, since the prosody of the parents' speech is matched by the body motions of the infant (Condon & Sander 1974). This synchronization is related to the alignment of verbal utterances found in dialogues (Linell 2009). It has been suggested that synchronization and alignment are the result of mirror neurons in the brain, and that it links the interactants and paves the way to human communication (e.g. Arbib 2002).

Another important socializing behaviour is the mutual gaze. Studies on gaze behaviour demonstrate that human infants engage in *mutual gaze* before the age of three months (Trevarthen 1979), and are able to follow someone's gaze and focus on the same object around the age of six months (Butterworth 1995). *Joint attention* is a prerequisite for word learning – in order to establish a connection between an object and a label the child must pay attention to the same phenomenon as the adult. This is achieved by gaze following or hand pointing. Pointing is a later development than gaze, and more tricky. If an adult points with the hand, the youngest infants tend to look at the out-stretched hand instead of the direction of the hand (see Chapter 4).

Also other primates use mutual gaze in mother-infant interaction. Chimpanzees display this behaviour at about the same degree as in human mother-child interaction (Bard et al. 2005). However, as with human groups, there are differences among chimpanzee groups. When comparing chimpanzees from two research institutes, one in Japan and one in the US, Bard et al. (2005) found that the Japanese group had significantly more mutual gaze than the US group. The gaze behaviour was related to the amount of tactile contact behaviour – the more the infants were cradled the less gaze behaviour took place (similar resuts have been found for humans, see LaVelli & Fogel 2002). The type of communication where the human infant checks back on the adults' face has not been observed in young chimpanzees. When finding new objects, they wait for the mother to manipulate it first and look at how she does it instead of searching eye contact (Tomonaga 2006).

1.4 Summary

This introductory chapter gives a brief overview of communication in humans and other animals, and points to some of the methodological problems that are involved in comparisons across species. Many of the issues that have been under debate during the last 100 years such as symbolic/referential function, combinatory features and learning are introduced and discussed. The understanding of the development processes in the individual is important in order to understand how communication works. Language is not there from the beginning, but the child acquires it by accommodating to others. Similarly, other species develop their respective communicative systems. Growing up in a social group means learning species-specific conventions, whom to groom and when, which vocalizations signal alarms and which signal food sources.

Some of the striking similarities seen across species occur in the initial stages of development and are then gradually reduced during the individual's development. For example, the early perception of speech sounds is similar in newborn humans and in other species, but only as long as the human baby has not yet acquired the structure of speech sounds in the surrounding language. There are similarities as well as differences between humans and nonhumans, as demonstrated by many of the studies. Many of the features considered uniquely human are found also in animals.

1.5 Suggested readings

Burling, R. (2005). The Talking Ape. Oxford: Oxford University Press.
An accessible introduction to some of the topics in language evolution, this book takes a helicopter perspective and discusses theories of the evolution of human language.

Hauser, M. D. (1996). The Evolution of Communication. Cambridge, Mass.: MIT Press.
This is the most comprehensive account of communication in humans and nonhumans today, covering theories, research questions, methods and results from a large amount of studies in neurobiology, evolutionary biology, ethology, cognitive and developmental psychology, linguistics and anthropology.

Hrdy, S. B. (2009). Mothers and Others. The evolutionary origins of mutual understanding. Cambridge, Mass.: Harvard University Press.
This book presents a new hypothesis of human evolution, suggesting that we have cooperative breeding practices to thank for a lot of human achievements, including language and culture. The book gives an overview of child-rearing practices in different parts of the world.

Chapter 2

Overview of biological signal systems

2.1 Introduction

Most animals have some form of communication, at least to find somebody to mate with. For animals living in social groups, communication is indispensable for recognizing group members, to find and share food, to warn others for dangers, and comprehend the warnings from others. The variation in communicative systems in the animal kingdom is enormous and the system for each species is highly developed and efficient. This chapter presents some of the different communicative signal systems based on the sense used to perceive the signal.

Communication can take place by sounds, movements, postures, touch, scents or electricity and it is received by the sense organs. Although many communicative expressions involve more than one sense, it is common to categorize the signal systems with respect to the senses involved, and, for the sake of simplicity, one sense at a time. The major categories are:

Acoustic signals	signals that are audible
Visual signals	signals that are visible
Chemical signals	olfactory signals, involving chemicals
Tactile signals	signals received by tactile sensory organs
Seismic signals	signals that are received through vibrations
Electrical signals	signals requiring electric organs

In this chapter, we will draw attention to communicative behaviours of some species that are not discussed in the individual Chapters 5–7 (primates, dogs and birds). We will present communicative behaviours in such different species as honeybees, dolphins, whales, and some domesticated animals such as sheep and horses. These species are chosen in order to show some of the diversity within the area of animal communication. At the same time, the choices illustrate some of the tendencies during the last 50 years in the field. The study of social behaviour in honeybees goes back to the 1950s with the well-known studies on the direction dance in bees, by the Nobel laureate Karl von Frisch (1954). During the following decades, studies of echolocation and its use for communication in dolphins and other whales reached a peak (e.g. Tavolga 1964, Payne 1983, Tyack 1983, 1986, Au 1993). More recently, there is an interest in communicative behaviour in domesticated animals such as goats, sheep and horses (Mills & McDonnell 2005, Sèbe et al. 2007).

2.2 Research methods and observational techniques

Communicative behaviour is investigated in observational/descriptive studies and in experimental studies. A descriptive study is often the first step. The traditional research cycle starts with observations of a phenomenon, and then proceeds to formulation of hypotheses, which are tested in experiments. After that, new observational data may be needed, new hypotheses and new experiments. In experiments, it is possible to control and manipulate the variables. For example, it can be tested whether a specific sound is communicative by playing it back to individuals of the same species and studying their reactions. The advantage with controlled experiments is that they can be carried out over and over again and it is possible to generalize from the results.

Human communication can be studied in different contexts, for example in the family, at the pre-school, school and work. What sets the study of human communication apart from studies of communication in other species is that it can be studied both by *observed* behaviour and by *reported* behaviour, where people are asked about what they would say or do in a given situation.

The settings for animal investigations may be the wild, zoos, aquariums or other places for captive animals, or laboratories. Studies in the wild are time-consuming, since the animals often have to be habituated to the human presence. Doing research on animals in captivity, on the other hand, restricts the chances of finding true natural behaviour. In many modern zoos, however, the animals live under conditions that are similar to what can be expected in the wild, with large enclosures and rich opportunities of social life. In laboratory settings there are possibilities to use some kind of surgical procedures, for example testosterone-implantation or deafening of songbirds to study the sensitive phase for song acquisition (Wada et al. 2004, Cynx & Clark 1998). Studies on captive animals, whether in zoos or in laboratories, always have the drawback that we never know whether the behaviour in captivity is the same as in nature. On the other hand, behaviours occurring in captivity but not seen in the wild may be indicative of underlying behavioural capacities (Candland & Bush 1995).

2.2.1 Playback as a method for investigating communication

Playback is maybe the most important tool in studies of animal communication – in the field as well as in a zoo or a laboratory. The first study with some kind of controlled acoustic playback was Regen (1913), who used a telephone line to study whether a female cricket at the one end of the line reacted to a male cricket calling at the other end. With the development of tape recording technology it became possible not only to send signals but also to record a call and play it back, and the playback method was invented. The playback method has been used successfully in many different species, insects as well as mammals and birds, for example wax moth (*Achroia grisella*; Jang and Greenfield, 1996), great tit (*Parus major*; Peake et al. 2005), domestic chicken (*Gallus gallus domesticus*; Evans et al. 1993), vervet monkey (*Cercopithecus aethiops johnstoni*; Struhsaker 1967), humpback whale (*Megaptera novaeangliae*; Tyack 1983), and elephant (*Loxodonta Africana*; Poole 1999).

Visual playback is not used as much as acoustic playback. One kind of visual playback was used in Tinbergen (1960) in his study of the black-headed gull (*Larus ridibundus*). After having made the observation that the black mask is most prominent during early spring and then disappears during the summer season, Tinbergen constructed different models of cotton to elicit reactions from the birds, and was able to decide which features were most important. Also studies using silhouettes of different shapes (for example a model of a hawk) have been successful in eliciting a predicted behaviour (cf. Chapter 7.2.2).

The playback method has also been used to test significance of chemical signals, for example by presenting samples of odour to subjects, so-called "sprayback" (Stevens 1975). Olfactory playback has been used for a wide range of species, for example rabbit (*Oryctolagus cuniculus*; Moncomble et al. 2005), snapping shrimp (*Alpheus heterochaelis*; Hughes 1996), spider (*Agelenopsis aperta*; Papke et al. 2001). Another version of the playback technique is the use of mechanical vibrations. This has been done for example to simulate synthetic male vibrations in the web by the spider *Cupiensius salei* (Schüch & Barth 1990), and to recruit honeybees (*Apis mellifera*) by simulating a foraging dance (Michelsen 1999).

Playback has been particularly useful in the investigation of referential signals, i.e. to investigate whether the signal in question encodes specific information, for example which kind of predator is approaching. Without the playback technique it would have been much more problematic to make such a claim.

2.2.2 Advances in technology influence the methods

There is a large range of different technical ways to document communicative behaviour, such as photographs, video recordings, by geophone (for seismic communication – developed for military use in Vietnam, Hill 2001) or hydrophone equipment (to study echolocation, also first developed for military use). However, it is important to keep in mind that the technology used restricts the findings, and 'you find what you look for'. To give one example, using hydrophones that are fixed to a wall in the dolphinarium restricts the possibilities to find out whether the dolphin vocalizations are directed towards another individual or not, whereas by attaching an acoustic tag *on* a dolphin directional vocalizations may be detected (Blomqvist 2004). This method would not have been developed without a hypothesis that the sound may be directional. It is evident that new questions together with new technology open up for new research results. During the last 30 years there has been a lot of development on sonar equipment to record and analyze echolocation clicks. These contain frequencies between 3 kHz and 150 kHz and are possible to record only by highly advanced technique. In the area of neurology, studies of neurological correlates to human language started with brain-damaged patients. Today, it is possible to investigate brain activities for example by techniques such as event-related fMRI (functional Magnetic Resonance Imaging) both in animals and humans (Gallese et al. 1996, Stamenov & Gallese 2002).

Instrumental analyses of vocalizations were first done with spectrograms, but advances in technology have made it possible to display the sound components on a computer screen, where they can be analyzed with different software programs. This increases the possibilities to build data bases of sounds and compare detailed analyses between different sounds, as well as between different individuals emitting the same type of e.g. alarm calls. Before these inventions it was customary to illustrate vocalizations by onomatopoetic (or sound-imitating) principles, for example using verbs such as *squeak, whistle, grunt, groan, moan, bark, purr*, and *clic* or to try to render the sounds by some kind of paralinguistic transcription, like *aarr, uh, eee* and *aaoo*. These descriptive labels can be seen as shorthand approaches to the discussion of units of communicative behaviour, and they are useful in fieldwork

2.3 Acoustic communication

Some kind of acoustic communication is used among most animals, but it is among birds and mammals that the most elaborated acoustic signals have been described. Sounds can be varied in different ways; for example, frequency spectrum, volume, duration, repetition and rhythm. They have an advantage in that they can be turned on and off instantaneously. However, since sounds may spread in all directions from the source it is difficult to control the signal so that it reaches only the intended receiver. The signal can in many cases reach the wrong receiver, for example a predator, which makes this channel sometimes risky to use. Some signals are graded, i.e. the strength of the signal is proportional to the strength of cause, for example the food call by chickens, which is produced in a higher rate for highly preferred food. Other signals have discrete characteristics, for example the sound combinations found in some birds (Catchpole & Slater 1995), primates (Züberbühler 2002) and whales (Payne & McVay 1971). Frogs and crickets send out sound pulses in species-specific series, similar to human Morse-signals (Manning & Dawkins 1992: 60).

2.3.1 When and where?

Acoustic signals are context-independent, meaning that they can be used anywhere and anytime. This is important for social animals living in areas where they may not be able to stay in visual contact with each other at all times. Sheep, for example, keep in contact with their group members by bleating sounds. The vocalizations are used from birth to maintain contact between the ewe and her lambs (Sèbe et al. 2007). The ewe can guide the behaviour of the offspring by changing the character of the bleat. Young lambs respond to high-pitched bleats from the mother by moving towards her, and to low-pitched bleats by staying close (Terrazas et al. 2002).

The way acoustic signals are used for early offspring-parent recognition varies between species. As mentioned in Chapter 1, the Nile crocodile has no individual call, but the calls from the young change in structure, so that the parent can always decide the age of the offspring and thus their need of protection.

Elephants have a large repertoire of acoustic signals, some of which are at low frequencies and not audible to humans, so-called subsonic communication. Elephants of both sexes use acoustic signals to advertise their reproductive readiness. The females have very short periods of oestrus: for the African elephant it is a matter of around 5 days every 4 years. They prefer breeding with males during the period when the males are in musth, which is a physiological condition with increased levels of testosterone and aggressive behaviour, lasting for up to 4 months. Since the female elephant may be miles away from a male in musth it is of vital importance to have an efficient way to communicate her oestrus state over long distances in order to find someone to mate with. The so called oestrus call is a sequence of rumbles that are audible up to four kilometres away, starting at low frequencies and then rising with high overtones, a pattern that is particularly efficient at long distances (Langbauer 2000). Female elephants are also engaged in antiphonal rumbling, i.e. calling in alternate and partly overlapping sequences. In a study of female captive elephants, Soltis et al. (2005a, b), found that five out of six were more likely to vocalize shortly after a rumble from another individual. Interestingly, the propensity to answer a rumble correlated to number of years spent together, so that of all the calls that were answered, 70% came from an affiliated female partner. This shows that vocalizations in elephants play an important role in social bonding.

2.3.2 How? Production and perception

Acoustic signals can be produced in several different ways. In most birds and mammals, for example humans, wolves (*Canis lupus*) and birds (*Aves*), sounds are created by air pressed from the respiratory organs. Whales and dolphins (*Cetaceans*) use a system of air canals and thin membranes in the nose where the air can circulate and be re-used instead of being blown out. The rattlesnake (*Crotalus atrox*) creates sound by vibrating the end of the tail, the rattle. Sounds can also be created when body parts are rubbed against each other, like when crickets (*Orthoptera*) communicate by rubbing their legs against their wings.

The frequency of sound is usually measured in cycles per second, or Hertz (Hz), i.e. the number of times the sound vibrates per second. Large bodies vibrate with larger oscillations and give low frequencies, while small bodies create high pitch tones. Moreover, the range of frequencies a species can hear generally correspond to their body size. Larger animals usually perceive lower frequency sounds than humans do, while small animals can hear sounds with higher frequencies. However, frequency ranges are also adapted to the typical environment. Low frequency sounds can travel longer distances and through dense rainforests, and therefore they are used by many solitary animals (e.g. Mack & Jones 2003).

An animal's hearing within different frequency ranges is adapted to the frequencies that are used in the species' communication. Humans are most sensitive to sounds between 100 and 3000 Hz, which correspond to the frequencies used in our speech. The sound that has been most important for us is, in other words, not the approaching predator or the escaping prey, but instead our own species-specific sound, the spoken language. We are able to hear sounds between around 16 Hz and 20 000 Hz, but there is a lot of communication that goes on, that we are unable to hear. Elephants, for example, use infrasounds (or subsonic calls), i.e. low frequencies below our hearing range and dolphins use sounds above our hearing range. Dolphins use frequencies up to 150 000 Hz. Dogs are able to perceive sounds up to 50 000 Hz. Specific whistles with high frequency sounds that are audible to dogs, but not to humans, have been developed for use in dog training. A grasshopper can hear frequencies of 50 000 Hz while moths hear sounds from about 1000 Hz up to extreme frequencies of 150 000 Hz. Moths can therefore perceive the ultrasounds that bats send out when they hunt, which gives them a chance to escape an attack. There is a large variation in the hearing abilities of different species, depending on the different design of hearing organs in the animal world and on the part of the nerve system and brain that is reserved to receive and analyze sounds.

In order to understand how sounds are perceived, it can be useful to start with what happens in human conversations. The speaker starts speaking by letting air pass through the vocal cords to create sound waves. These sound waves are spread from the mouth and nose in all directions. Some of the sound waves reach the eardrums, the middle ear, the inner ear with the cochlea, and are changed into nerve impulses that the brain can perceive. All vertebrates have ears designed after the same basic plan as ours but there are many modifications to adapt them to the needs of each particular species.

Insects do not react to sounds in exactly the same way that we do. Our eardrum is constructed in a way that the sound waves can only hit the part of the membrane facing outwards. In many insects the sound waves are transmitted through air canals or the whole body so that also the backside of the eardrum take in vibrations. Other insects hear with sensory hairs or antennas specially designed to sense the vibrating air. The acoustic communication of fish seems to be most pronounced during the breeding season. Many fish can grind their teeth and this sound is amplified by the resonance of the swim bladder. Fish like the eel create sound by pressing air bubbles through the mouth while other fish produce sounds with the swim bladder. Since fish lack outer ears and cochlea, it was long assumed that they were not able to hear at all. When the Nobel Prize laureate Karl von Frisch reported that fish could hear, this was denied by the German scientists, and he started to work on honey bees instead (Gould & Gould 1994: 90). Later, however, his findings on fish hearing were acknowledged and today it is agreed that some fish have very good hearing. The hearing organs in fish vary in design but they are usually relatively simple and adapted to receive signals in the water rather than the air. The swim bladder is often involved in sound reception and works as an acoustic amplifier. Water-living mammals like whales and seals have ears very similar to our own. These animals originate from terrestrial mammals, which have returned to the water and the ears have gone through changes so that they fulfil the needs in the marine environment.

2.3.3 Echolocation and other sounds in dolphins and whales

The notion of a silent sea world was radically challenged when researchers like Schevill and Lawrence (1949) reported lively underwater acoustic communication, in particular in animals belonging to the order *Cetacea* (dolphins and whales). This finding, and the availability of new technology in form of hydrophones, triggered an outburst of studies during the following decades (for overviews, see for example Tavolga 1964, Payne 1983, Au 1993). The sounds produced by cetaceans seem to serve different functions: echolocation clicks are typically used for orientation and locating prey, whereas other sounds are used for social communication. Echolocation is a unique way of reading the environment since the same individual is both sender and receiver of the message (it is therefore sometimes called "auto-communication", Bradbury & Vehrencamp 1998, Sebeok 1972). The animal produces short pulses and obtains information from the returning echo. The echoes that are received give information about the surrounding environment such as distance to obstacles and prey. Echolocation can be seen as a complement to vision, and it is a practical tool when visibility is poor, like in deep water, or in darkness. The clicks are emitted through a rounded organ (the melon) in the forehead. The incoming echoes pass through by the lower jaw to the ear and the hearing centre in the brain. Signals with low frequencies (25 Hz) travel long distances and are used to scan the environment. Higher frequencies (up to 120 000 Hz) are used for example when the animal approaches an object and wants more detailed information about the structure of the object. For example, a fish may be recognized through the air in the swim bladders.

Research on cetaceans has focused mainly on dolphins, particularly on the bottlenose dolphin (*Tursiops truncatus*). Many dolphin species live in a complex social organisation of fission-fusion character, i.e. the group composition changes over time (see also primates, Chapter 5, and parrots, Chapter 7). When visibility is restricted, the communication is mainly based on acoustic signals described as whistles, clicks, grunts, barks, cries, squeaks, chirps and quacks (Poulter 1968). A lot of research has dealt with cognitive aspects and the ability of dolphins to understand and learn human language (e.g. Herman et al. 1984, 1999, Reiss et al. 1997). The results from these experimental studies have demonstrated that dolphins are able to comprehend referential pointing (Herman et al. 1999), they understand the concept of non-existence (Herman & Forestall 1985), they can understand a primitive grammar with linear order (Herman et al. 1984), and moreover, they can imitate human-designed whistles (Reiss & McCowan 1993). This shows that dolphins have great cognitive capacities and they have a talent for vocal learning, used for example in signature matching, i.e. when they accommodate to each other (Janik 2000). One reason for the dolphin's understanding of referential communication could be that this is part of their natural behaviour. In echolocation the dolphin is oriented in the direction of the beam. "The whole body of the dolphin as well as the emitted sonar field are thus, in a sense, pointing toward an object" (Herman et al. 1999: 363, see also Au 1993).

Traditionally, it has been assumed that clicks are used for navigation whereas other sounds are used for social communication in dolphins. However, it is also possible to categorize the sounds used for communicative purposes into two types, whistles and broadband pulses (Blomqvist 2004). Whistles are the sounds that have attracted the most interest (in particular the so-called signature whistles, 94% of the whistles in dolphins, according to Caldwell & Caldwell 1990). The whistle serves as an individual identification marker and is used to establish and maintain contact (Sayigh et al. 1998, Tyack 1986). Since the social group is characterized by constant separations and reunions and dolphins travel over vast distances it is imperative to have a reliable way to recognize each other and to keep in contact. Mothers and small infants generally travel together, but even small infants often move around on their own and need to have a safe means for identification. Another sound used for social communication is the broadband pulse. As mentioned above, Blomqvist and colleagues (Blomqvist 2004, Blomqvist & Amundin 2004 a, b) were able to identify aggressive pulses directly aimed towards another dolphin by using a specifically designed acoustic tag.

The sound was labelled the "Machine-gun" sound because of its repetition rate. The behavioural response of the received pulse sound was commonly an escalation of the fight with other and additional aggressive displays, for example forceful head movements and jawclasps, i.e. opening and forceful closure of the mouth, resulting in a loud acoustic "bang". The use of an aggressive pulse sound, having directional properties, directed at a specific opponent indicate a resemblance to the provocative fixed visual staring, which is common in many land living mammal species (Blomqvist 2004).

Vocal learning plays an important role in dolphin communication. Some features of the learning process, such as babbling and overproduction, suggest interesting parallels between development in dolphins, songbirds, nonhuman and human primates (Elowson et al. 1998, McCowan et al. 1999, McCowan & Reiss 2001). A number of studies have investigated the gradual development of whistles, showing that infants rarely whistle when together with the mother, but whistle a lot when separated. Two functions of the signal have been suggested: the young dolphins whistle to tell their position and they whistle to regulate the mother's behaviour: to make the mother wait, or to change her position so that the infant can get in "infant position" to rest and to nurse. Thus, at the same time as the whistles help infants to keep in contact with the mother, they also give freedom to the infants to move around in the water (Smolker et al. 1993, Tyack 1998). A benefit of vocal learning is that the whistle is an individual signature and functions as kin recognition, preventing mating of closely related individuals. Sayigh et al. (1990) studied mother-calf pairs and discovered that the whistle went through a series of gradual modifications and were stabilized at the age of about one year.

Studies on larger whales differ in perspective from research on dolphins, due to both practical problems in keeping captive whales for experiments, and to the problems involved in identifying individuals in the oceans. Instead of training individuals to understand human acoustic and manual signals, the research has concentrated around analyses of their communicative behaviour, particularly of the structures of the songs (e.g. Payne & McVay 1971). The functions are still rather unclear, but it is assumed that the sounds are used as advertisement displays, in pair formation, in courting, and possibly, as a kind of food calls. The most studied species are humpback whales (*Megaptera novaeangliae*), killer whales (*Orcinus orca*), and sperm whales (*Physeter macrocephalus*). Different whale species lead very different social lives; for example, humpback whales, like the dolphins, live in fission-fusion societies, whereas killer whales mainly stay in their matrilineal group, and the sperm whales are organized with females in matrilineal groups (with the calves suckling up to the age of 10–15 years) and adult males leading solitary lives (Rendell & Whitehead 2001). Like dolphins, whales use both echolocation and other sounds. Vocal learning seems to be important. For example, in the complex songs of humpback whales vocal convergence in the group has been discerned. The themes of the songs change over the years, and therefore it has been suggested that these whales have cultures, similar to humans, and that their songs are culturally transmitted (Eriksen et al. 2005, Noad et al. 2000). The simultaneous change of the songs in many places at the same time may be facilitated by the use of oceanic deep sound channel (Rendell & Whitehead 2001). Also in the killer whales, vocal learning and dialectal features in the calls have been found (e.g. Ford & Fisher 1983, Yurk et al. 2002). Each group has a dialect, or group-specific call repertoires. Individuals in neighbouring groups may use some of the calls, but not the entire repertoire. The function of this variation has been suggested to be that of group cohesion. "These group-specific call repertoires in killer whales are thought to indicate pod affiliation, maintain pod coherence, and to coordinate activities of pod members" (Tyack 1998:199). Killer whales have also been found to use socializing whistles during close-range interactions (Thomsen et al. 2002). For the sperm whales, where the adult males live solitary lives, less is known about the vocalizations. However also these whales are known to have some kind of group dialect, developed by social learning (Rendell & Whitehead 2001, 2004). Thus, vocal learning seems to be common in most whale species. But which parts are learned and which parts are innate?

Spectrographic analyses make it possible to identify units at different levels. In a groundbreaking analysis of humpback whale songs, Payne and McVay (1971) recognized the following six recurring elements: subunit < unit < phrase < theme < song < song session. The elements could differ in their precise configurations, but the sequences were always ordered in the same way (A, B, C, D, E and not A, B, D, C, E). This motivated a classification of the humpback whale vocalizations as songs: "Humpback whales (*Megaptera novaeangliae*) produce a series of beautiful and varied sounds for a period of 7 to 30 minutes and then repeat the same series with considerable precision. We call such a performance 'singing' and each repeated series of sounds a 'song'." (Payne & McVay 1971:597). Like in analyses of bird song, the sequential order of units in humpback whale vocalizations has been discussed in terms of syntactic patterns, and compared to human speech. We will return to this issue in Chapter 8.

2.4 Visual communication

Visual signals can be produced in many different ways and they vary along two dimensions. On the one hand, a visual signal can, for example, consist of colour contrasts, ornaments, and body postures. But another dimension is the potential that is given by movement. Colours may be changed, covered and uncovered, body size and posture may change rapidly, for example when turkeys, peacocks and lyrebirds fan their tails. When a deer erects its tail, the white markings become visible and this warns other individuals of a danger. Some other visual signals involve rhythmic movements, for example the fiddler crabs waving their claws to attract females, or the dance displays by many birds, for example grouses and cranes.

An advantage with temporary visual signals is that they can be turned on and off very quickly which makes them easy to direct to the right receiver. The disadvantage, however, is that the signal may not be discovered by the intended receiver unless the receiver is relatively nearby and the view is clear.

2.4.1 When and where?

Animals using visual signals usually communicate during daytime because they are dependent on sunlight. Fireflies (*Pteroptyx tener*) belong to one of the few species communicating with visual signals during the night since they themselves produce light. The males produce a species-specific flashing pattern of light pulses and the females respond with another blinking pattern and this facilitates the flies' search for partners. The best environment in which to use visual signals is in open surroundings. Complex visual signals operate best on short distances and at larger distances the signal must be very simple and not involve too much information. The distance also depends on the size of the signalling animal, and visual signals are typically used in courtships and fights when animals are close to each other.

Visual communication can be performed for example by colour patterns, body postures and movements, facial expressions and ornaments of different kinds.

2.4.2 How? Visual capacities

What we see depends on how the light is absorbed or reflected by objects. The part of the spectrum that we usually call light consists of wavelengths of 400–700 nanometres (400–700 "million parts" of a millimetre). Eyes of other animals are often sensitive to other wavelengths. Many insects and birds have developed an ability to perceive ultraviolet radiation and they probably see the colours of flowers in a different way than we do. Rattlesnakes have special "pit organs" that perceive infrared radiation given from living bodies. By using this "thermal image" of the living prey they are able to strike and kill even if they are blind (Kardong & Mackessy 1991). Rattlesnakes are assumed to use both vision and their facial pits when striking a prey.

The field of vision also differs between species. One prominent difference is that between predator and prey. In humans, the eyes are placed on the front part of the head, which is the same position as in cats, wolves, eagles and other hunters. Prey species like sheep, rabbits, hen and ducks on the other hand need to see as much as possible of their surroundings in order to discover predators and they have one eye on each side of the head. Most of these animals have a blind area behind the head, but the mallard duck (*Anas platyrhynchos*) has a full panoramic vision, that is a 360-degree field of vision. This allows them a good chance to detect predators approaching from the back (Martin 1986).

2.4.3 Colour patterns

Colour signalling is common in fish, and it is particularly conspicuous in coral reef fish. In some fish ultraviolet colorations have been found (e.g. Losey 2003, Siebeck 2004). Since many of the predators in this environment do not have UV vision these signals can be used as a secret language among conspecifics, without risk of being noticed by predators. The function is often connected to reproductive behaviour, but change of body colour may also signal submission to avoid aggressive conflicts (e.g. O'Connor et al. 1999). Also butterflies and birds are known to have spectacular colouring and particular visual capacities. Butterflies are sensitive to polarized light, and they are able to discriminate between objects from the angle of polarization (Sweeney et al. 2003). Bird colour vision differs from vision in humans in that birds have a broader spectral range, involving also near-ultraviolet light.

The colours of some birds are striking and attractive to a human eye, but the difference in visual capacities between human and birds makes it difficult for us to estimate the exact visual signalling in birds. For example, it is quite possible that there are colour patterns in what we perceive as dull and uninteresting birds, since we are not able to detect how they look. ".. to study these traits from the narrow perspective of human colour vision is to miss the true diversity of colours seen through avian eyes and is, in fact erroneous." (Cuthill et al. 2000:75).

2.4.4 Body postures and facial expressions

The body posture can give important information to other individuals. Generally, a large body is a sign of strength and to pretend to have a larger body is a common way to gain dominance. A high ranked wolf walks with stiff legs in an upright posture and high tail, while submissive individuals minimize their body size (e.g. Schenkel 1967). The tail is an important part in visual communication, and it is often possible to interpret feelings and intentions just by looking at the tail. In many species, tail postures and tail movements are involved in social contacts like greetings, threats and courtship displays.

Facial expressions can carry a lot of information. For example the ears, nostrils, lips and eyes are useful. Most research on facial expressions has concentrated on primates, most likely because their expressions are so similar to human facial expressions and our nonverbal communication. We share many grimaces, tongue movements and eye movements with other primates. With the faces human and nonhuman primates can express fear, anger, curiosity and indifference (more about this in Chapters 4–5).

Social animals living on open grasslands often use visual signals rather than acoustic because they are very rarely out of sight, but may be too far away for sound. A social animal like the horse, for example, has a well-developed body language. It is easy to see a horse's mood simply by looking at its ears. Ears pointing forward show that the horse is alert and interested while ears lying flat backwards show aggression or fear. Apart from the ear postures an attacking horse may open its mouth, and draw the corners of the mouth backwards, and show the teeth.

2.5 Tactile communication

For many species, tactile communication is the first type of communication with newborn infants. Humans and chimpanzees hold, kiss and cuddle the babies, horses, cows and sheep lick the offspring. Tactile communication is assumed to have an appeasing effect. It often takes place in intimate contexts and is harder to investigate than for example acoustic and visual communication. However, absence of touch can be conspicuous for example when people that do not know each other are gathering together in a queue, in a waiting room or in an elevator.

2.5.1 When and where?

Appeasement and reconciliation signals are often tactile, and they are found to reduce tension and heart rate (Feh & de Mazières 1993). It is common to use tactile signals in greetings. Many animals greet each other with touching, buffing carefully and smelling each other to show recognition. Chimpanzees and humans hug and embrace, and wolves do jaw wrestling and mouth licking. When elephants greet each other they touch each other's mouths with the tips of their trunks. Lions and many other cats greet by rubbing their heads and foreheads against each other.

A certain form of tactile signals available in both invertebrates and vertebrates is allogrooming, i.e. cleaning the fur or skin of another individual. The allogrooming behaviour is assumed to fulfil several different functions at the same time: improve hygiene, strengthen social bonds and decrease aggressiveness in the group. It has been suggested that allogrooming has its origin in taking care of the young (Lazaro-Perea et al. 2004). Another function of grooming is a premating mechanism (Stopka & Graciasová 2001). It is generally the dominant individuals that are being groomed by lower-ranked individuals, but sometimes a high-ranked individual may show reconciliation by grooming a low-ranked individual. An advantage when using tactile signals it that one can be sure that the signal reaches the intended receiver.

Horses are often involved in social grooming. Two horses stand parallel to each other with their heads pointing in opposite directions. The grooming consists of careful bites along the neck and all over the back. Usually a horse only has one or a few "grooming-partners" in the flock. Grooming at the preferred place – the base of the neck – has shown to effectively reduce heart rate both in foals and in adult horses (Feh & de Mazières 1993).

2.5.2 How? Skin, whiskers and sidelines

Tactile signals are received by pressure detectors and/or by hair cells on the body. A human body is covered with 1.5–2 square meters of skin. This surface is more than a thousand times larger than the light-sensitive retina in both eyes together and ten thousand times larger than the total surface of our two eardrums. Despite this, our skin is mostly used for regulation of body temperature and not for communication. For other species the body surface may be the most important receptor of communicative signals. Whiskers (or *vibrissae*) are specialized sensory structures that transmit tactile information to the brain. When the whiskers are bent at the base every hair can perceive both direction and speed of the bending. Whiskers are important for foraging in many marine mammals, for example seal and walrus. The picture below illustrates the whiskers, placed in rows around the snout, in walrus (*Odobenus rosmarus*).

Vibrissae in walrus (Odobenus rosmarus). Photo: Jennie Westander

Fish have a sideline system with organs consisting of clusters of sensitive hairs, which are bent by the movement of the water. Each organ will receive different signals and this makes it possible for the fish to localize the source. The sideline system also makes it easier for the fish to keep its place in the shoal.

2.6 Indirect tactile signals – the seismic signal system

Vibrations cover the need of communication for many different kinds of animals. A spider web, for example, functions like a telegraph wire. Male spiders stimulate a potential partner by pulling the web. Spiders have a cannibalistic nature and it is important for the male to prevent this behaviour in the female. If the female responds by running towards the vibrations, the male will have a warning that she is hungry and he will run away. However, if the female merely signals that she is there the male will advance a few steps at a time to approach her, stopping now and then to shake the web. Vibrating the web is assumed to stimulate the female to take on a mating position, and at the same time inhibit her predatory tendency (Maklakov et al. 2003). The jumping spider (*Habronattus dossenus*) combines seismic signals with visual signals in courtship behaviour, making the courtship more complex. The jumping spider uses both a zigzag display of colourful body parts and seismic signals of vibration (Elias et al. 2005).

Some animals communicate by creating vibrations in the ground or on the water surface. They use seismic signals, or substrate vibration signals. This type of signals can be transported longer than for example acoustic signals, which makes them suitable for communication over long distances. The white-lipped frog (*Leptodactylus albilabris*) that lives in Puerto Rico's rainforest is a species that is extremely sensitive to seismic signals. It tucks into the mud and expands its air bag to the muddy surface, producing a seismic drum that can be perceived by other frogs in the surroundings (Lewis & Narins 1985). Another interesting example is the Kangaroo Rat (*Dipodomys spectabilis*), which marks its territory by drumming with the foot and thereby sends seismic signals to other individuals. Randall & Matocq (1997) report that mothers drum more than nonmothers, which shows that the signal is probably directed at the offspring. Moles (*Chrysochloriadae*) communicate by drumming the hind legs on the floor in the subterranean passages (Narins et al. 1997, Hill 2001) or by banging their head against the walls (Klauer et al. 1997). These vibrations may be perceived by other moles, both in the same passage and in adjacent passages. Moles have special ear ossicles to hear through bone instead of through the hearing pathway, and they put their jaws against the walls to hear better. Vision is so superfluous in a life under ground that the eyes of the mole have been covered with skin and fur during the evolution. Finally, animals that live on the water surface can communicate by hitting the water to create vibrations or small waves. An example is the pond-skater (*Gerris remigris*). The signals give information about the sex of the signalling animal and are used for courtship communication or defence of a territory (Wilcox 1979).

2.7 Chemical communication

Judging from the focus of research interest, chemical communication is particularly important in reproduction (e.g. elephants, Langbauer 2000; humans, Pause 2004; spiders, Maklakov et al. 2003; tamarins, Heymann 2000). However, chemical cues are also important in individual recognition, kin recognition, territory marking, and marking of a food source (Miller et al. 2003). As in other senses, every animal species has its own spectrum of smell. They are most sensitive to compounds important to their lifestyle. Rabbits can smell a carrot or a dandelion while cats or dogs do not care about such smells but instead are interested in the smell of the rabbit. Research on chemical communication took a new turn when pheromones were identified (from Greek *pherein* 'to transfer' and *hormon* 'to excite'). Pheromones are defined as "substances which are secreted to the outside by an individual and received by a second individual of the same species, in which they release a specific reaction, for example, a definite behaviour or a developmental process" (Karlsson & Lüscher 1959: 55). Communication with pheromones has since then been registered in insects, fish, amphibians, reptiles, rodents, marsupials, felines, canids and primates. Most of what is known about pheromones comes from studies of insects' chemical communication, however, probably because of the large economic problems those insects cause in agriculture. There is a practical interest in knowing how we can use insect communication to stop noxious insects from destroying the harvest. Furthermore, chemical compounds used by insects are usually relatively simple and easy to synthesize which facilitates the research.

Chemical signals are divided into two groups: *pheromones*, which are used between individuals of the same species and received by some kind of smelling organ, and *allomones*, which are detected by other species as well, and often in direct contact with the source, e.g. by taste. Pheromones typically activate sexual behaviour, but they are also used to mark territory, signal alarm and recognize group members.

2.7.1 When and where?

Chemical signals can be used to communicate on large distances and to leave messages (similarly to human writings) that remain even when the animal has left the area. This makes them suitable when visual and acoustic signals can be difficult to perceive like in darkness or brushy environments, and in solitary animals that do not often meet conspecifics. It is also the ideal signal system for communication over long distances under water. But the system is not only useful at long distances. For a more precise identification the chemical signals are used at short distances. If a dog smells something interesting, it will try to come as close as possible to the source of the smell, not stopping until it is a matter of millimetres.

The chemical signal system is slow, which may be a disadvantage. It may take long time to change the message because of the long fading time and during that time there is a risk to announce one's presence to e.g. a rival, or a predator. Furthermore, it is difficult to direct chemical signals to a certain receiver because they are spread with the wind or water currents. Insects differ from other species in having a system of pheromones, which only some receivers can detect. In other animals the chemical signals are not that elaborated, but different mammals for example can learn to recognize each other's odours. In horses the olfactory signal has shown to be a much better identification marker than acoustic signals (Rubenstein & Hack 1992).

2.7.2 How?

How a chemical signal is spread depends on what species is involved. Many animals have scent glands that produce an individual-specific secretion with chemical substances. A common way to distribute the signals is by rubbing and stroking the glands against trees and other objects and also other individuals. Glands can be found on several different places on the body and are usually most concentrated around the genital and anal areas. To give examples of some different gland placements, the Tasmanian devil, the badger, and the guinea pig have glands around the anus, the reindeer on the inside of the hock, and the hyena and the wolf between the pads. The elephant has a gland behind the eyes, where a liquid is produced when "in musk" i.e. when they are ready to mate.

Many animals have several glands, positioned at different places. Domestic cats have glands between the eye and the ear, around the lips, under the chin, on the back, and around the tail. The deer has glands in the corner of the eyes, between the digits, outside and inside the hock and around the anus. Humans have glands on palms and soles, at the scalp and forehead, but also a specific scent gland located in the armpit. This gland placement is unique to humans, gorillas and chimpanzees and is assumed to have developed as a result of the upright posture. The effect of the odour is enhanced by the growth of hair in the armpit.

The glands in some mammals have connection to pockets or sacks, which makes it easier to spread the secretion and leave smell tracks. Animals that do not have so many glands send chemical signals through faeces, urine and genitals. Rhinos are not very territorial but to frighten intruders and avoid conflicts they mark their territories with piles of faeces around the territory grounds. Sometimes several rhinos in the same area use the same pile and they scratch around in the pile to spread their own smell. Many animals try to place their markings as high as possible, while the hippo spreads its urine by wagging the tail when urinating, and male cats point their urine backwards.

The characteristics of chemical signals depend on the chemical composition of the pheromones. Large molecules with high molecule weight are present for a longer time since they are less volatile and these are therefore often used for territorial markings. Pheromones can also function as warning signals. If an animal is hurt, pheromones are released and these warn other animals of the danger. For example, when an earthworm warns other individuals small molecules with low molecular weight are produced and spread easily. Flighty molecules have short duration, which makes the signal disappear quickly after the danger is gone.

Around the territory of many mammals there is often a kind of invisible olfactory fence. An example of this is the badger, which uses the glands at the tail root to mark stones, tree trunks and even the ground on chosen places in the territory by pressing its back hard against the object. Such markings work like chemical nametags and give information about territory owner. Chemical markings can also be used as demonstrations of strength and ranking.

Ants are highly eusocial (i.e. have a caste system with division of labour) and they protect and defend their nest. There is a strong need for the individuals to recognize each other and ants are well known for an excellent ability to discriminate between nest mates and strangers. It has been suggested that the identification is based both on genetic and environmental factors. Pheromones are important in ant communication and there are two different kinds of pheromones. One is a "colony odour" that helps the ants to differentiate between ants that belong to the colony and strangers. Some ants mark the surrounding territory with pheromones to keep strangers out. Since the territorial marking keeps other ants away these ants do not have the same need for nest mate recognition. The other group of pheromones are alarm pheromones. As with bee alarm signals, these signal immediate attack, and working ants stop with what they are doing to help in the attack instead.

In large colonies with a clear division of labour, aggression within the group is uncommon. In ant species without morphological differences between queen and workers there is more of intraspecific aggression. Heinze et al. (1999) report on a study of the myrmicine ant *Eutetramorium mocquersi*, from Madagascar, where intense fighting between individuals was found. The threatening ant shows a stilting posture with inflated gasters, which makes the impression that the ant is larger than the real size. Since the size of the working ants increases with increasing colony size, so that a large ant is a sign of an ant from a large colony, it is possible for an ant not only to pretend to be strong but also to feign a larger colony by adopting a dominance posture.

Since our own sense of smell is quite underdeveloped in comparison with other animals' the importance of smell in communication has long been a rather unexplored area and knowledge is still limited. We need special equipment to register chemical signals and to recognize communication with chemical signals.

2.7.3 The olfactory system

The main olfactory system consists of sensory neurons on both sides of the olfactory bulb in the nasal cavity. When we breathe air is pulled in and reaches the olfactory epithelia. The human epithelia covers an area of about three-square centimetres, which is very small, compared to many other mammals'. Most cats have for example epithelia covering about 14 square centimetres and in a dog it can be as large as 75 square centimetres. In addition to this, amphibians, reptiles and most mammals have a special sensory organ, namely the vomeronasal organ or Jacobson's organ. It is usually located beneath the nasal cavity and has a direct contact with hypothalamus instead of having the nerve connections to the olfactory system. The vomeronasal organ does not have any connection to consciousness and therefore the signals can influence behaviours without the individual being consciously aware of it. Snakes catch molecules with their tongue and taste them by inserting the tongue into the vomeronasal organ. Mammals need direct contact with chemical substances for example by licking or nosing on things in their surroundings, which makes the molecules attach to the tongue or nose. To get the molecules to reach the vomeronasal organ some animals, for example cattle, deer, zebras, horses and elephants, display a *flehmen* behaviour. The lips are lifted, the nostrils are closed and the head is turned from side to side a few times while the animal takes a deep breath. The elephant uses the trunk to sniff the scent and puts the tip of the trunk directly into the vomeronasal organ positioned in the roof of the mouth. The observation of flehmen in an animal has shown to be of practical use when it comes to detecting oestrus and improve reproduction (Rekwot et al. 2001, Sankar & Archunan 2004).

The existence and function of a vomeronasal organ in humans is still unclear and has been under debate for some time. It has long been assumed to be vestigial, i.e. present in the fetus but not functional (e.g. Albone 1984). New findings have led researchers to other interpretations, for example that the human vomeronasal organ is reorganized during the first weeks (Tirindelli et al. 1998), that it is degenerated with pseudogenes (Brennan & Keverne 2004) or that it is so different from the vomeronasal organ in most other mammals (except for chimpanzees) that it has not been possible to identify it (Smith et al. 2002). This discussion has implications for the question about human pheromones, since pheromone reception is generally seen as a function of the vomeronasal organ, even though there is also evidence indicating that pheromones can be received by the main olfactory system (e.g. Dorries et al. 1997).

2.8 Electrical communication

Electric communication is mainly known in fish. The electric signals are produced in special electric organs. When the signal is discharged the electric organ will be negatively loaded compared to the head and an electric field is created around the fish. A weak electric current is created also in ordinary muscle cells when they contract. In the electric organ the muscle cells are connected in larger chunks, which makes the total current intensity larger than in ordinary muscles. The fish varies the signals by changing the form of the electric field or the frequency of discharging. The system is only working over small distances, about one to two meters. This is an advantage since the species using the signal system often live in large groups with several other species. If many fish send out signals at the same time, the short range decrease the risk of interference.

Electric fish can be divided into weakly and strongly electric. The weakly electric fish, e.g. the South American knife fish (*Gymnotus carapo*) and the African elephant nose fish (*Gnathonemus petersi*), emit electric signals of a few volts only. They use these signals for orientation (so called electrolocation) and for social communication. The males court females by sending out certain frequencies of electrical signals. From these, the female can read the size and social status of the male and also the male's engagement in courtship behaviour. Experiments have shown that females prefer males with long bodies, which correlate with the amplitude of the electric signal (Curtis & Stoddard 2003, Triefenback & Zakon 2003). The strongly electric fish, such as the electric eel (*Electrophorus electricus*) from South America, and fish of the *Torpedinidae* family can produce signals of up to 900 volts. This may paralyse and kill prey and also be dangerous to humans.

2.9 The multimodal honeybee – acoustic, visual, chemical, seismic communication

The most widely known example of an advanced communicative system is the dance of the honeybee (*Apis mellifera*), first described in detail by Karl von Frisch (1954, 1967), as expressed through both vibrations and touch. Like for example ants, bees are eusocial animals, i.e. they live in tight colonies and rely on a high degree of cooperation. All labour is strictly divided. Workers process and store food and the young are taken care of collectively. There are morphological differences between queens and workers, so that it is only the queen who is able to reproduce. This strict division of labour reduces the amount of nest conflicts. There is no need for signals of dominance and submission, but what needs to be communicated is mainly information about food sources, nest defence and member identification.

The communication about food sources in honeybees has long been seen as closely related to the human verbal language because the bee is not expressing an inner emotional state but conveying information about something that is outside the immediate situation. In other words, they use the referential function.

A bee that has found a food source returns to the hive and releases a "recruitment pheromone" which has a strong effect on the other bees. Inside the hive, the bee starts to dance. The function of this dance is to inform the other members of the colony that there is a food supply somewhere and also to indicate the place of the food. In order to investigate the details of the dance, von Frisch trained marked bees to find specific odours at different distances from the hive. What he found was that the bees use alternative dances, depending on the distance to the food source. If the food source is close to the hive (closer than 100 meters) they use a round dance, and if it is further away, they use the so-called waggle dance. In the *round dance* the bee is running in small circles to the right and left. The more often a bee changes direction in the dance, the richer is the food source. Foraging bees in the hive are attracted by the dance and follow the dancer, touching her body with their antennas. After the forager has left the hive the newly recruited bees also leave, and search around the hive. Presumably, the recruiters find the nearby source by remembering the odour from the dancer. If the food source is further away, the information communicated is more complex, as it contains details not only about the kind of food, but also about distance and direction to the food source. This information is conveyed in the *waggle dance*. In this dance, the bee moves in semicircles interrupted by a short running straightforward when wagging her body from side to side, making a buzzing sound. The procedure is repeated to the right and to the left, so that the two semicircles form one circle with the straight run in the middle. The distance to the food is translated into the duration of the straight run. The slower the dance is performed the farther away is the food. In fact, it is not the geographical distance that is indicated, but rather the amount of energy connected with the distance. For example the distance is indicated as longer if there is a strong wind.

The direction to the food source is shown in the angle of the straight waggle part in the dance. Bees use a so-called geocentric system, i.e. the location is defined by the axis of rotation of the earth, and not in relation to the own position and body axis (the egocentric method, which is preferred in many human cultures). The geocentric system makes it possible for the bee to translate the angle of the food source to the sun to a certain angle to the horizontal surface inside the hive. As the sun angle changes during the day, the bee also changes the dance angle. When the information is received the other bees in the hive can fly out to find the food, helped also by the odour from the dancing bee.

The recruiting dance seems to be innate. However, the receivers do not only react instinctively to the dance, but they also rely on learned knowledge, for example landmarks in the environment. Gould (1990) designed a study where he trained foragers to find pollen placed in a boat in a lake. When they were allowed back to the hive, they danced without success: the other bees refused to believe that there could be a food source in the middle of a lake, and the dancing foragers did not manage to recruit other bees. However, when the boat was placed at the shore of the lake the response was more positive. Even when the boat was placed at the far end of the lake and the bees had to fly over the water, they followed the forager. Gould's explanation is that the bees in the experiment relied on their mental maps: since they had never found pollen in a lake before, they did not venture a trip there. At the shore it was more probable that there could be pollen.

The honeybee dance is a good example of multimodal communication, and also how new methods can add new information about animal communication. The hypothesis first put forward by von Frisch was that the waggle dance was related to the angle to the sun, and that the bees were using by seismic and olfactory channels. Later research has shown that more channels are used. By using an artificial model of a bee, covered with bee wax and vibrating electrically, Michelsen (1999) was able to demonstrate that sounds also play an important role. Without the acoustic part, the bees did not follow the model's instructions, but when a moving wing was added to the model, producing an acoustic sound, the other bees followed.

Even if the recruiting is successful and the other bees will leave the hive to forage, the hive will not be totally deserted. The workers will stay and prepare for the processing of the food that will soon be arriving, brought home by the foragers. Since there are two completely different reactions to the dance, Seeley (1992) suggests that the dance in the hive in fact can be said to have two different meanings. For the workers in the hive it means that they should prepare for more food processing, for the foragers it means that they should abandon their search for other food sources but instead follow the dancer. In order to ensure that no intruders enter the hive, which the owning bees have invested so much work in, they have to guard and defend the entrance. Strangers are recognized by their flight pattern or by their odour. Bees in the same colony acquire the same odour because of the relative proportions of food from different flowers. However, other bees in the same area may acquire similar odours, and mutual robbery among neighbours is common. If intruders are detected however, the alarm system starts. The alarm consists of a specific alarm pheromone, which has an activating effect on all bees in or near the hive. A typical behaviour of aroused bees is to fly or run around searching for somebody to attack. Since the attack is well synchronized, an intruder can be surrounded by hundreds of bees within seconds.

2.10 Summary

This chapter has presented some methodological aspects of studies on communication. Most research on animal communication implies focussing on only one sense, or communicative channel at a time. However, this is more for practical purposes, since multimodal communication in reality is the most common. The acoustic signals are easiest for humans to notice, record, analyze, and describe, whereas it can be harder to register visual, tactile and chemical signals. One problem is to identify whether signals are communicative signals or not. The most efficient method to date is to use playback technique to control for the validity of a signal. The different communicative channels place different demands on the environment. The acoustic channel can be used in a variety of situations and settings, day and night, at long and short distances, whereas the visual signals are restricted to good light and closer distances. Chemical signals are efficient in darkness, under the ground, and in brushy vegetation. They are important in species that lead a solitary life and don't know when their signal will be picked up by a conspecific animal. The characteristics of each communicative channel are summarized in Table 2.1 below.

Table 2.1 Communicative expressions (after Bradbury & Vehrencamp 1998: 567)

Channel	Medium requirements	Maximum range	Localizability	Signal duration
Auditory	Air or water	Large	Medium	Short
Visual	Ambient light	Medium-large	Good	Variable
Tactile	None	Short	Good	Short
Chemical	Current flow, air/water	Large	Variable	Long
Electric	Water	Short	Good	Short

2.11 Suggested readings

Hauser, M. & Konishi, M. (Eds.). (1999). The design of animal communication. Cambridge. Mass.: MIT Press.
   Hauser and Konishi use Tinbergen's four perspectives as a point of departure to describe the communicative systems in a broad range of species. They discuss underlying mechanisms of behaviour and how behaviours have evolved differently in different species.

Maynard Smith, J. & Harper, D. (2003). Animal Signals. Oxford. Oxford University Press.
   This book looks into different theoretical standpoints and terminology. It gives a good overview of the diversity in signal systems of different species as well as the problems involved in creating a coherent framework for the analysis of animal signals.

Wyatt, T. D. (2003). Pheromones and Animal Behaviour. Communicating by Smell and Taste. Cambridge. Cambridge University Press.
In this book the reader gets a summary of chemical communication. The chapters cover production and perception, pheromones for reproduction, alarm, territory marking, orientation etc. It is written in a readerfriendly style, accessible also for non-chemists.

Chapter 3

Human language – its use and learning

Human verbal (and nonverbal) communication

3.1 Introduction

The situation above illustrates what is maybe the most basic of human activities – a group of people engaged in talking. As soon as two or more human beings get together, talk and gestures materialize: we converse, chat, gossip, discuss, argue, and quarrel. We ask about what has happened, we talk about people we know and people we have heard about, we talk about what they say and think. Through language we are able to move in time and space, discuss past events and speculate about the future, disclose beliefs and dreams. While we talk, we use gestures, and the other interactants give verbal (*Really?*) and nonverbal (nodding) feed-back. For a long time, linguistic research has focussed on monologic language use and language used in written texts and we know less about the dynamics in dialogues with two or more participants. However, in the context of *speech alignment*, Pickering and Garrod (2006) find a way to describe the contributions of the participants in a dialogue as a joint activity "just as two dancers' movements constitute a joint activity of ballroom dancing" (Pickering & Garrod 2006: 220). The participants do their best to follow, cooperate and align their contributions. Empirical data has shown that there is a tendency for speakers to reuse the same words and constructions that have been used earlier by other speakers in a conversation (e.g. Szmrecsanyi 2005). In this chapter, we will come back to alignment in dialogues when describing language acquisition by children and adults. But first some words about language use in a more general sense.

Language is a conventionalized code, agreed upon in the speech society, which each human child spends the first years of life in learning. The basis is the use of words: spoken, written or signed words, which are joined together by grammatical rules. As far as we know, this way of communicating makes us different from other animals. It is most often used in the acoustic medium, but also visual and tactile channels are used, for example in writing and sign languages. In this chapter, however, the focus will be on spoken language, i. e the acoustic channel. Before we go into how infants discover and develop language we will give a brief account of language functions and forms around the world.

3.1.1 Social life and the languages of the world

There is an enormous variation in human social life, evidenced in the many cultural and religious traditions, natural sources, political systems – and languages. It is a prominent feature of the human language that every group has its own version, with words and structures that are not understood by the neighbouring groups. One's own language is often perceived as the most natural way of communicating, and the languages of others may be seen with suspicion. For example, the Kayardild-speaking group in Australia calls their mother tongue the 'strong language' whereas the neighbours' language is called 'mumbling talk' (Evans 2010: 51). This practice highlights the fact that language is an in-group phenomenon, based on implicit agreements between the group members – they are the ones that form and maintain the language. Language works as the glue that holds a group together and keeps outsiders out.

Generally, speakers use the dialect, or language, that was used in the region where they grew up, but they also have a potential to change when coming into contact with speakers with other dialects, or languages. Speech accommodation to others reflects an ambition to show increased solidarity at the same time as it is a result of positive contacts. Typically, it is revealed in the synchronization of speech habits, for example speech rate, pronunciation of certain sounds, or use of in-group jargon (Giles 1980, Giles et al. 1973).

Languages are not equally distributed over the world, but they occupy different ecological niches, often separated by natural barriers such as mountains or swamps. An approximate estimation of the current situation tells us that there are 6000–7000 languages spoken in the world. Some of these languages are spoken by hundreds of millions of speakers, whereas other are only used by small groups. Most languages are found in Asia and Africa, whereas Europe is the continent exhibiting the lowest number of languages.

In the spirit of Linneaus, the languages of the world are divided into language families, with assumed genetic relationship. Their exact number of depends on how detailed the description is. At the website www.ethnologue.com (November 18th 2012) 116 different language families are listed. Six families dominate; these cover two-thirds of all languages and are spoken by 5/6 of the world's population. These families are Afroasiatic and Niger-Congo (both spoken in Africa), Austronesian (spoken in South East Asia), Sino-Tibetan (spoken in East and south East Asia), Trans New Guinea (spoken in the Pacific), and Indo-European. Indo-European languages (eg. English, French, Russian, Greek, Persian, Hindi) are spoken as first or second languages throughout the world. The Indo-European family dominates in numbers of speakers (almost half of the world's population speaks an Indo-European language) but not in numbers of languages. The smallest figure for number of speakers is found in the family Trans New Guinea, with approximately 500 languages and less than 1% of the world's population. This part of the world has a reputation among linguists for having an unusually rich repertoire of languages. The explanation is a social system with a high value placed on multilingualism. For example, in Gapun in Papua New Guinea, it is not uncommon for people to know four or five languages, at least not for the older generation (Kulick 1992).

If a child grows up monolingually or bilingually is basically a matter of the political situation in the environment. The monolingual situation is often a result of official language planning by law regulation. This is evidenced by the fact that many so-called traditional cultures without language planning have bilingualism or multilingualism as the normal case. Linguistic exogamy (i.e. to marry outside one's language group), which is the practice in many traditional societies in Africa, Asia and Aboriginal Australia, involves bi- or multilingualism. Being multilingual enables one to have multiple social identities within a community, and to keep up a complex contact network. As mentioned earlier, in Papua New Guinea, a country with many languages and a small population, multilingualism is a highly valued asset. The traditional society regarded language as an important boundary-marker, at the same time as all groups were equal in importance (Kulick 1992). There are obvious practical advantages in speaking more than one language; it gives more contacts and increased access to information. For example, knowing the meaning behind local names of flowers and herbs gives an advantage when it comes to using them for medical treatment. In many European cultures, access to education was only made possible through the learning of Latin during the Middle Age. Today it is English that dominates as an international language, and in many places people have to become bilingual in order to get access to higher education.

India is an interesting example of a multilingual country with bilingualism on both the societal and the individual level. The official language is Hindi, written in Devanagari script, with English (Latin alphabet) as an associate official language. In addition, there are more than 1600 local and regional languages used across the country. India's multilingual nature becomes clear when one looks at the text of the bank notes. The front side of a ten-rupee note contains the words "ten rupees" in Hindi and English, and on the reverse side there is a language panel of fifteen languages written in ten different scripts (the languages are: Assamese, Bengali, Gujarati, Kannada, Kashmiri, Konkani, Malayalam, Marathi, Nepali, Oriya, Punjabi, Sanskrit, Tamil, Telugu, and Urdu). In the constitution given in 1961, a Three-Language-Formula was suggested,

Multilingual road sign in New Delhi, India. Photo: Jennie Westander

based on the idea that each individual needs three languages: one for identity with the local group, one for the national Indian identity, and one for international communication (the Indian prime minister Nehru referred to English as 'the international window'). In contrast to the attitudes in many Western countries, bilingualism is seen as a positive feature, something to be proud over: "We have an edge over monolingual countries in that we acquire two or three or sometimes four languages with ease, and use them for a variety of purposes" (Mathew 1997:165).

Road signs illustrate how multilingualism constitutes a salient part of the linguistic landscape. The road sign on previous page is written in four languages, with four different scripts; Hindi (in Devanagari script) English (Latin script), Punjabi (Gurkmukhi script), and Urdu (Perso-Arabic script). The languages are important for identity – speakers of the four languages all get a feeling of belonging.

3.1.2 Observation techniques

The study of the human verbal language involves a number of different techniques, depending on the aspect of language that is to be investigated. *Sociolinguists* and linguists within anthropology study language in social settings, how people go about interacting in daily life. This methodology parallels field studies in animal research, aiming at observing natural social behaviour. One difference is that the linguist may also ask informants about which communicative behaviour is appropriate in certain situations, thus adding an introspective facet to the observation. In studies based on participant observation, the investigator gets to know the members of the speech community personally and plays an active role in the social group. A different method is to set up experiments and place people in situations where they are asked to perform certain activities, for example to address a stranger or ask for help. *Psycholinguists* are interested in language in the individual, how humans process, produce and perceive language, how languages are acquired, and how languages may be forgotten. The data may be observational, for example with recordings of speech, or experimental, where informants perform certain activities. Research on *production* and *perception* of language may involve different kinds of technology. As mentioned in Chapter 2, acoustic aspects of language production can be analysed by sound spectrograms where frequency and intensity of the sounds are displayed. By using speech synthesisers it is possible to control the variables when investigating for example how people perceive different speech sounds. *Neurolinguists* focus on activities in the brain, for example which regions in the brain that are active in certain linguistic tasks. Development within electrophysiology and neuroimaging techniques has 'opened the window to the brain', in the sense that it is possible to study the brain in detail. The cerebral blood flow may be measured by PET-scanning (positron emission tomography) or fMRI (functional magnetic resonance imagery). The grey matter substance is assessed by sMRI (structural magnetic resonance imagery). By measuring the electrical activity in the brain through ERPs (event-related brain potentials), it is possible to study the reaction of the brain to speech stimuli in a time resolution of milliseconds. For example, an unexpected sentence structure may be visible on the computer screen as a "silent protest" from the brain. In combination with behavioural observations these methods give us new insights about correlates to language functions in the brain.

Studies of language development in children have a long history, starting with descriptions, where adults take daily notes of the child's communicative behaviour (often their own children, see for example Darwin 1877). When recording technology became available in the 1960s, it was possible to increase the database and collect data from more children. There are two main designs: longitudinal studies, where individual children are recorded with regular intervals for a longer period of time, and cross-sectional studies, where data are collected from groups of children of different ages. The longitudinal project led by Roger Brown (reported on in his 1973 book, see below) represents a leap forward in language acquisition research, and it involved many of the researchers who were later to dominate the field. Another great leap forward was the establishment of the CHILDES database by Brian MacWhinney (CHIld Language Data Exchange System, MacWhinney 2000). This database contains transcripts and audio files from children of a large number of different languages, together with devices for automatic analysis of the data.

There are many different ways to assess perception and recognition of speech in children. For the very young infant, before the age of two months, the high-amplitude sucking paradigm is often used. This method uses sucking rate as a measure of speech discrimination. Any new phenomenon in the speech, for example a change from [pa] to [ba], will elicit more sucking. When the infant is able to sit up, the Headturn Preference Procedure (also called Preferential Looking Paradigm) may be used. The infant sits in the caregiver's lap and orients towards one of two screens, namely the one that matches the linguistic stimuli. As we will describe below, the Headturn Preference Procedure is often used for testing of word comprehension. More recently, the same methods to measure electrical brain signals in adults (see above) have been used also in young children.

In order to describe structural aspects, for example the grammar or words of a particular language, either introspection (self-examination) or direct observation may be used. Introspection is possible when the researcher speaks the language in question and is able to decide which constructions are appropriate in which situations. Introspection is often combined with observation, where for example speech production is recorded and analyzed. Another method is to use questionnaires, for example where informants are asked to judge given examples.

3.1.3 Language functions

Even if we agree on language as a tool for communication it can be used in more than one way. It is common to distinguish three communicative functions that are used in daily life: (1) the referential function; (2) the social function, and (3) the affective function (see more on functions in Jakobson 1960, Halliday 1975, Robinson 1972). As mentioned earlier, the referential function has often been assumed to be *the* function, exclusive to human language, whereas social and affective communicative functions are shared among humans and nonhumans. This is a simplification, however, since utterances are generally multifunctional, and the same utterance can contain several functions. For example, a simple statement like *It is raining today* may be interpreted as strictly referential and give information about weather conditions, for example in telephone conversations when the interlocutor does not have the same weather experience. However, it can also serve a social function, which is obvious in situations when the parties are in the same setting and able to observe the same weather. Then talking about rain does not add any new information, but serves as social glue. Furthermore, the utterance *It is raining today* can also have an affective function, expressing personal engagement and disappointment of rainy weather.

Box 3.1 Three communicative functions of language (after Jakobson 1960)

1. **Referential function.** By this function we are able to refer to external events and objects, and to place them in time and space by words such as *here, there, in the city, now, then, before, next week*. This is traditionally considered to be the hallmark of human language. Two subtypes can be distinguished; the declarative (a statement such as "I bought a dictionary") and the interrogative ("Did you buy anything in town?"). The referential function is what is used in most writing.

2. **Social function.** A social utterance serves to establish communication, to check that everyone is on friendly terms and that the communication channel works, or the appropriate language is used. For example, certain rituals are often used in greetings and partings ("Hi, how are you?" – "Fine, and you?"). It can be compared to the wagging of the tail in dogs. In many cultures it is customary to comment on the weather.

3. **Affective function.** This function differs from the others in that the speaker is sometimes not addressing a listener but only expressing inner emotions. Illustrative examples may be the kind of vocalizations used when missing a bus, dropping a glass, winning at the lottery.

Besides these everyday functions, there are more specific language functions for particular circumstances. The British philosopher J. L. Austin (1962) observed that certain utterances do not communicate events, but they perform acts which can change the whole situation. Austin uses the label *performatives*. For example, if somebody in the right position says *I pronounce you man and wife* to a couple, the couple is married, and if someone says *I promise to be there* the promise is made. Utterances like these are often expressed by fixed structures, with first-person as subject and a verb in present tense, and they sometimes include a word corresponding to the English *hereby*. Performative utterances stand in sharp contrast to other utterances in that they don't have a truth-value; they cannot be true or false (you cannot say *No you don't!* if someone says *I pronounce you man and wife*, or *I promise to come*).

3.2 The child's discovery of language – the first year of life

Children use the affective and social language functions, before they start using the referential function. The British linguist Michael Halliday (1975) argues that communicative functions can be identified as early as before the emergence of the first words. Analyzing data from his own son Nigel, Halliday found that the social and affective functions were present before the age of one year, whereas the referential function, informing others about something, was not used until Nigel approached the age of two years. This can be related to the child's cognitive development. Since young infants are not aware of the fact that others may not have access to the same information that the child has, they consequently don't see a need to inform them about facts that the child knows.

In order to use language to communicate with the social group, each sound, word and grammatical rule of the particular language has to be learned. How can young children achieve such a complex task? And how come young children in fact are better language learners than adults are? In this section, we will describe human language from the perspective of the young child's discovery, and let the child guide us through different linguistics levels; sounds, words, and grammar.

3.2.1 Parent-infant interaction

As soon as the baby is born, it is immersed in intense verbal and nonverbal interaction. Not only are the adults around the baby constantly talking, cuddling and smiling but also the baby takes active part in the interaction. Already in the first week of life, infants have been found to mirror the adult facial expressions, and move arms and legs in synchrony with the parent's speech rhythm (Condon & Sander 1974, Trevarthen 1977, Meltzoff 1981). We will present examples of these nonverbal behaviours in next chapter, Chapter 4.

In many parts of the world, infants are treated as conversational partners already from birth. Parents, other adults and older children spend much time with the baby, talking, chanting, and singing in a very special manner, which is different from what is used in conversations with adults. The particular way of interacting has been labelled *motherese, infant-directed speech* or *Baby Talk* (Ferguson 1977). It can be seen as a special realization of alignment in that the adult tries to decrease the distance between the child's vocalizations and the adult way of pronunciation. The most prominent characteristic of Baby Talk is a high pitch, and abrupt pitch changes up and down. This seems to attract the child's attention. In studies where fundamental frequency is measured, statistically significant differences have been found between speech directed to an adult and speech directed to infants (e.g. Garnica 1977). The fundamental frequency is higher in child-directed speech. Perhaps this is as a function of the so-called 'frequency code' (Morton 1977, Ohala 1984). According to this, low-pitched vocalizations are used to express threat and high-pitched vocalizations to express the opposite. Another characteristic is a hyperarticulation of certain speech sounds. For example, the three vowels /i/, /u/ and /a/ (as in sheep, shoe and shark), placed at the corners of the vowel triangle, are more extended as adults address infants, which results in an expanded vowel space (Kuhl et al. 1997, Burnham et al. 2002). Also grammatical structures are adjusted. Sentences are short, subordination is not used, and verbs are used in present tense. The lexicon is small, with repetitions of the same words over and over again, and sometimes names are used instead of first and second person pronouns (*Mummy will help Carl*, instead of *I will help you*).

Baby Talk has been assumed to be beneficial for language acquisition, helping the child analyze the linguistic input (Garnica 1977, Newport et al. 1977; for an interesting experiment with computer modelling, see de Boer 2005). However, also speech addressed to dogs contains these modifications (Burnham et al. 2002, cf. Chapter 6.8), which means that it probably has social functions as well.

However, cross-cultural studies show that the language acquisition process is robust enough to also happen without Baby Talk. For example, in families in Guatemala, Samoa, Papua New Guinea, Java and Inuit communities in Canada, children seem to observe others rather than being involved in verbal interaction and eye contact (Ochs 1988, Pye 1991, Kulick 1992, Crago 1992, Gaskins 2006). In these cultures, the children are not expected to interact, and the adults do not use the typical pattern with high and variable pitch. Pye (1991) describes the situation for children learning K'iche', a Mayan language (Box 3.2).

Despite cultural differences in the details of interaction practices, all children spend their lives surrounded by others, and they acquire their first language in the

> **Box 3.2** Parent-child interaction in Guatemala
>
> **CLIFTON PYE REPORTS ON PARENT-CHILD INTERACTION IN GUATEMALA**
>
> Parent-child interaction is very different from the American middleclass standard. K'iche' babies are kept close to their mothers at all times, either strapped to their back, in a craddle of rags nearby, or beside them in bed. The mothers are quick to interpret any movement or vocalization as a signal to feed their babies, which they can do while continuing with their own activities. They will also quiet a baby they are carrying on their backs by gently rocking forwards and backwards while patting it on the bottom and saying "sh, sh, sh…" in a soft voice (*kukux ka' chila'*). Occasionally a mother will amuse her baby with her necklace, flowers, or bits of string, but for the most part babies are ignored. (Pye, C. 1991:235)

same way. They align to others and manage to pick up enough of the surrounding language to be able to produce meaningful words around the age of one year, and grammatical utterances at two-three years of age. At this age they already surpass what animals are able to do with human language, even those animals that are exposed to careful training and teaching of a language.

3.2.2 The "little universalist" – early perception of speech sounds

Infants' speech perception shows sensitivity towards human speech structures already from start. Infants appear to be tuned into listening to human voices and work with "grouping and sorting the range of utterances to which they are exposed" (Jusczyk 1997:107). Children are known to be able to discriminate between speech sounds long before they can produce them, and they discriminate most in the earliest stages of life. Using the method of measuring high-amplitude sucking rate, Eimas and colleagues (Eimas et al. 1971) tested 1-month and 4-month old infants on stimuli from synthetic speech. The results demonstrated that even the 1-month old infants were able to discriminate between the voiced consonant /b/ and the unvoiced consonant /p/. Since this behaviour was found to occur in young infants it was suggested that categorical perception was "part of the biological makeup of the organism" (Eimas et al. 1971:306) and that "children come equipped with these skills" (Pinker 1994:264). As mentioned in Chapter 1, however, as similar results are reported also for nonhumans, it can be concluded that "uniquely human processing is not essential for these particular tasks" (Kuhl & Miller 1975:72). Of course, this result only tells us that the ability is not specifically human, it does not question the importance of early discriminating abilities in human children. Later studies have showed that birth is not even the starting point for auditory hearing and learning in children. DeCasper and colleagues asked pregnant women to recite particular passages before the child was born, and found reactions showing recognition both before and after birth (DeCasper et al. 1994).

A number of studies have shown that infants are more sensitive to speech sounds during the first month of life than they will ever be during the rest of their lives. They are able to perceive contrasts that are not part of the surrounding language, a talent which is lost during the first year. The contrast between [ra] and [la], which Japanese-speaking children are able to discriminate at the age of 6–8 months, is lost around the age of 10–12 months (Kuhl 1999). Similarly, the ability of English-speaking children to discriminate between certain contrasts made in Zulu, at the age of 6–8 months, is also lost around the age 10–12 months (Best et al. 1995). Their acoustic cues have now been shaped by the linguistic experiences (Iversen et al. 2003). It seems plausible that the change in perception is due to children starting to acquire meaning in language. "This change in sensitivity to phonetic contrasts occurs not simply from lack or presence of exposure, but from exposure to phonetic contrasts that are used to contrast meaning in the native language" (Werker 1995:103). What happens is that infants use their growing experience of the surrounding language (or languages) to sort out the language-specific phoneme contrasts. Thus, instead of having to learn new phonetic contrasts one by one, children gradually drop contrasts.

This dropping of contrasts could have been the end of the story of sound structure acquisition – but it is not. Both children and adults are able to reopen the system and to learn new contrasts in second language learning, even if they do not always reach total success, but have to end up with approximations (giving a foreign accent). Internationally adopted children, however, that do not have access to their first language any longer, but only their second language, have been found to lose their first contrasts totally (Pallier et al. 2003). Ventureyra et al. (2004) studied 18 Koreans that had been adopted to France between the ages of 3 years and 9 years. Nine of them had not been exposed to Korean and nine had been back to Korea. They were tested on six distinctions. For five of the distinctions there was no recognition at all by any of the children. One distinction could be identified by the group that had been back as tourists. This shows that contrasts that were once important can be lost in the absence of continuous exposure.

3.2.3 Early vocalization – babbling

Newborn children vocalize already from the beginning – they grunt, scream and cry. But it takes a while before they can use sound structures from the surrounding language. Human speaking demands control of a number of details; the breathing, the vocal cords in the larynx, the lips and the tongue. By quick movements of the tongue, the lips and the velum (the back part of the roof of the mouth) we create obstacles to the air stream to produce different speech sounds. It is therefore important that the tongue movements are precise.

In the earliest phases of speech production, children are unable to control their articulation in this way. Instead, what we hear are particular vocalizations called babbling. Babbling sounds astonishingly alike across children, and it is not possible to decide which language the children are targeting until after several months. This is partly due to the anatomical and physiological changes during the first year in life. Before three months of age, the human infant's vocal tract construction is in fact more like that of a nonhuman primate than that of an adult human, with very little space to articulate, and tongue movements specialized for sucking. At around three months of age, the larynx starts to sink and the vocal tract becomes more and more similar to adult humans. The modification of larynx and surrounding muscles continues during the first two-three years (Juszcyk 1997).

Several stages have been identified in the development towards pronunciation of words in the target language (e.g. Oller et al. 1999). The first time after birth is characterized by vowel-like sounds (quasivowels). After two to four months, 'cooing' sounds emerge. Next milestone is the expansion stage with "vocal play", when children seem to be experimenting with their voices, and gradually gain more control over their vocalizations. They use long sequences of vowels with varying tone contours. Around the age of six to nine months, there is an important developmental change, when rhythmic patterns emerge and syllabic babbling is produced (*bababa, gogogo*). Now infant vocalizations start sounding more "language-like", and the relative frequency of language-specific vowels and consonants increases. As a result, it is possible to differentiate different languages in the babble (Vihman et al. 1986, Boysson-Bardies et al. 1989, Roug et al. 1989). Deaf children vocalize in a similar way as hearing children during the first months, but they do not reach the stage of canonical syllabic babbling. The syllabic babbling is assumed to be an important step towards spoken language and it is also used as a diagnostic tool to decide language impairments. Thus, this first year of language development implies a gradual progress from universal to language-specific speech sounds – both in perception and in production. Box 3.3 summarizes some universal milestones for the first vocalizations in children, from the first week to 6–9 months.

Some of the sounds used in these early vocalizations disappear and are not used again when the child grows up. But there are also sounds that stay on into adulthood.

Box 3.3 Milestones for babbling (after Oller et al. 1999: 225)

1. Phonation stage: quasivowels, /aa/, glottals /mm/
2. Primitive articulation stage: cooing /ngaa/, /chuu/
3. Expansion stage: full vowels with consonants: /daa/, /bo/
4. Canonical syllabic stage: reduplicated sequences: /mamama/, /didi/, /bidibidi/

Grunts are a kind of glottals that emerge early in infants and remain in adult communication as feedback signal (*mm, uhuh*) or accompanying effort (*uuh*). McCune et al. (1996) see grunts as important to theories about language development – they give a link between infants and adults, but also between humans and other primates, since nonhuman primates, e.g. chimpanzees, vervet monkeys and gorillas, also use grunts.

For adults, it has been shown that there is an activation of tongue muscles when listening to speech. The tongue muscles match the sound produced by the speaker (Fadiga et al. 2002). This can be interpreted as a help in understanding others who share the same linguistic code and articulary motor repertoire. We do not yet know when this mirroring behaviour starts in the young infant.

3.3 Language in the toddler

After a year of listening to language and trying to identify recurring and meaningful sequences in the stream of speech in the environment, the child is ready to enter "the world of words". This implies that a particular form is connected to a particular meaning. In this process of mapping sounds to meanings, the infant is helped by adults' pointing at objects in the environment and labelling them (e.g. *there is a car*).

3.3.1 Mapping forms to meanings

Around the age of one year the first clear mappings between sounds and meaning emerge and the child's vocalizations can be interpreted as words. The canonical babbling stage, with *bababa, mamama, didi*, turns into the word stage in such a smooth way that it can be difficult to pinpoint exactly when the first word is used. It is no coincidence that many of the earliest words that are reported by parents resemble canonical babbling *mummy, daddy, kitty, doggie*. The interpretation of what infants are saying is influenced by the parents' expectations. Thus, many parents report *mummy* or *daddy* as the first words. This is not the expectation in all cultures, however. Kulick (1992) describes a situation in Papua New Guinea where parents see children as independent and strong individuals and expect them to want to leave the traditional Taiap culture. Their first words are reported to be *mnda* as in "I'm sick of this", and *aiata* as in "stop this" in Taiap.

The vocabulary grows slowly at the beginning, but a couple of months after the second birthday many children enter a period that has been called "vocabulary spurt" or "word explosion", when they learn up to ten new words a day. The speed of vocabulary growth has intrigued researchers, and a specific word learning mechanism "fast mapping", i.e. an ability to learn a word after only a few exposures has been suggested. This rapid word learning can be seen as a prerequisite for the increase in lexicon. The fast mapping phenomenon can be tested experimentally in the Headturn Preference

Procedure, where infants do not have to say anything but can display word knowledge by turning the head and look at a screen, after a short exposure. For example, the child is exposed to invented words e.g. *sarl* or *bard* together with unknown objects on the screen, then the words are presented again and the child's looking preference is registered (Schafer & Plunkett 1998, Houston-Price et al. 2005; for fast mappings in dogs see Chapter 6).

The acquisition of words by young children can serve as an illustrative example of how word meanings are defined. In dictionaries words are listed with defined meanings. This gives an impression of meanings as fixed and stable, which is a simplification of way words are used in communicative situations. Instead, words get their meanings in the context where they occur. Linell (2009) gives the example of different meanings of the word *new*. In the utterance *my new favorite philosopher* it means that the person has just discovered this philosopher, in *my new car* it indicates that it is a car the person just got, in *new car* said by a car salesperson *new* means that it is a new product. Linell suggests that meanings are sets of semantic resources that are "used in combination with contextual factors to prompt and give rise to situated meanings" (Linell 2009: 330). Children learn word meanings by meeting the word in many different contexts, and in the beginning the words may not mean the same to children and adults. In her "semantic feature hypothesis", Clark (1978) suggests that children start with only a few semantic features of a word and then gradually add more and more features. For example, if the child uses the word *doggie* not only for dogs, but also for cats, cows, horses it seems plausible that only the feature [four-legged] is used, resulting in over-extensions of the meaning of *doggie*. Later, the child may add the features [barks] and [wags tail] to doggie and [horns] and [udder] for cows, thereby discriminating them. Although there are many examples showing evidence for overextended uses of words (*daddy* for all men, *light* for both lamp and moon), there are also examples of children's errors that cannot be explained by this model. There might be an earlier period, when children learn the meaning of a word in a specific, frequently used, context and then the initial meaning is stretched to other contexts (Bowerman 1978). Thus, the meaning of a word may be underspecified in an early phase, e.g. *car* meaning only the family's car, and overextended at a later phase, when *car* means everything that makes a motor sound, to finally reach a meaning that is equivalent to the meaning used by most speakers in the specific speech community.

3.3.2 Cultural differences reflected in children's language

Related to the question of word learning is the question of cultural learning. Children are not only entering a world of word meanings, they are also entering the world. Learning language is part of the socialization process. This view has a strong promoter in the British linguist Michael Halliday.

> Learning one's mother tongue is learning the uses of language, and the meanings, or rather the meaning potential, associated with them. The structures, the words and the sounds are the realization of this meaning potential. Language learning is learning how to mean.
> (Halliday 1973: 24)

During the last decades there has been an increasing interest in finding out how children "learn how to mean" in different languages. A number of cross-linguistic studies have demonstrated both similarities and differences between children growing up with different languages. To take the similarities first, the words *mommy, daddy* and greeting words like *hi/hello* seem to be the first words for children in many cultures – at least as reported by parents.

However, many differences have been identified. In a study on "Baby's first words", involving more than 900 English-speaking, Cantonese-speaking, and Mandarin-speaking children around the age of 11 months, Tardif et al. (2008) found some interesting differences. English-speaking children tended to use more words depicting objects whereas the Mandarin-speaking children used more words relating to family members. Cantonese-speaking children were in between the other groups. The results reflect both cultural and linguistic differences – not only do Mandarin-speaking children talk more about family members, some of the labels used do not even exist in English. For example, in English there are no specific terms differentiating between older and younger sister, maternal and paternal grandparent, or maternal and paternal aunt. Below we give the top-ten words for each language, translated into English.

Table 3.1 The ten most commonly used words, as reported by parents, and percentage of children who produce them (Tardif et al. 2008)

English (USA n = 264)	Cantonese (Hong Kong n = 367)	Mandarin (Beijing n = 336)
Daddy (54%)	Aah (60%)	Mommy (87%)
Mommy (50%)	Mommy (57%)	Daddy (85%)
BaaBaa (33%)	Daddy (54%)	Grandma-paternal (40%)
Bye (25%)	YumYum (36%)	Grandpa-paternal (17%)
Hi (24%)	Sister-Older (21%)	Hello?/Wei? (14%)
UhOh (20%)	UhOh (Aiyou) (20%)	Hit (12%)
Grr (16%)	Hit (18%)	Uncle-paternal (11%)
Bottle (13%)	Hello/Wei (13%)	Grab/Grasp (9%)
YumYum (13%)	Milk (13%)	Auntie-maternal (8%)
Dog (12%)	Naughty (8%)	Bye (8%)

As is shown in the table, 2 out of 10 words are family names in English, and 3 out of 10 are family names in Cantonese. This is in contrast to the Mandarin-speaking children, who have 6 family names on the top-ten. This difference suggests that labelling family relations is more important in the socialization of Mandarin than in the other two languages.

Thus, we can conclude that early interaction practices differ between English-speaking and Mandarin-speaking children. But there are also more subtle differences between children of different languages, revealing that they are tuned into specific ways of thinking by acquiring a particular language.

To illustrate this point, we will give some examples of comparative studies with children learning different languages. In a much-quoted study of English-speaking and Korean-speaking children, Bowerman and Choi (2003) found that all children enjoy putting things into containers, and piling objects on top of each other, but the way they categorize the actions reveals that they have totally different concepts. In English, the preposition *on* is used both for putting a cup on the table and putting a top on a pen, whereas the preposition *in* is used to put a book into a box-cover. In Korean, however, the same verb (*kkita*) is used for putting top on pen and book in box-cover, whereas a different verb (*nohta*) is used for putting a cup on the table. This implies that Korean children make a distinction between putting things into tight-fit interlocking containers and putting them on a loose-fit surface, whereas English children distinguish between putting something into container or putting something on top of something. Such cross-linguistic differences in the way children talk may be found already around the age of two years (Bowerman & Choi 2003).

There are other examples about two-year-olds making culturally different distinctions. Children acquiring Tzeltal (a Mayan language spoken in Mexico) use a language-specific differentiation for words for eating, indicating whether it is something soft, *lo*, something crunchy, *kúx* or bread or tortilla, *wé* (Brown 2001), Samoan-speaking children differentiate between different kinds of coconut palms by the look of their frond (Kernan 1969), Turkish-speaking children distinguish between self-experienced or hear-say when they use a verb (Slobin & Aksu 1982).

After these examples of typological differences between the semantics of languages, and how children come to learn them, we would like to join Levinson and Wilkins in the following conclusion:

> The implications are that the child language learner is a constructivist – he or she is not just mapping local forms onto pre-existing innate concepts but building those concepts as he or she learns the language. (Levinson & Wilkins 2006: 551–552)

3.3.3 Words and world-views – what do you call your cousin?

As suggested above, our world-view is acquired together with the first language. How different can different world-views be? The arbitrary relation between content and expression implies that the same content can be expressed by different forms. The different words for 'horse' exemplified in Chapter 1, all stand for the same concept, and it is easy to understand that different languages have different words. However, not all words can be translated that easily. Just as different languages cut up the sound spectrum in different slices (for example by having a contrast between [r] and [l], or not), they also carve up the physical surroundings in different pieces by selecting and naming events and objects. Generally, when a certain concept is important in a culture and therefore there is a need to talk about it in a precise way, words are developed to express the concept. It was no coincidence that the Mandarin-speaking children above, who talked a lot about family members, also had particular words defining the exact nature of the relationship. Words tell us how different languages categorize the world, and thus also something about the culture and environment in that specific speech community. It does not say that the same content cannot be expressed in another language, only that when an area is important in a certain culture, distinct words are used, whereas the same content is expressed by longer phrases in other languages.

An area, which has attracted a lot of interest, is how family members are labelled, i.e. kinship terminology. The notion of marrying groups, or moieties, lies behind many kinship terminology systems. In many cultures the ideal person to marry is one's cross-cousin, i.e. a man should marry his mother's brother's daughter or his father's sister's daughter, but not his mother's sister's daughter or his father's brother's daughter. In order to keep track of family relations it is important to have an exact terminology – in some languages the relations are even grammaticalized with different pronouns for different relations (in Lardil, 'we' = 'me and my brother' is expressed differently from 'we' = 'me and my father', see Evans 2003).

Here, we will present in more detail the system in Warumungu, a language spoken in Central Australia (Simpson 2002). In Warumungu, as in many other languages, the same word may be used for different persons sharing a particular relationship, for example, a woman and her mother-in-law both call each other *ngunarri*, and a man and his mother-in-law refer to each other as *miyimi*. In English, the word *cousin* is used in this reciprocal way, but not mother-in-law and daughter-in law. Sometimes the same non-reciprocal term may refer to different persons. An English example is *brother-in-law*: the brother-in-law may be the sister's husband, or the wife's brother, but has to be a male. In Warumungu, however, the gender differentiation is not central. The word *kampaju* refers not only to father and father's brother, but also to father's sisters. The box below gives a overview of the terminology for the inner family, mother, father, sister, brother, siblings, aunt, uncle, cousins, grandmother and grandfather.

The number of words for the immediate family is roughly the same in both languages but there are some interesting differences as to how their meanings are distributed. The English word mother refers to only one person, the same with father. All the other English words may be used to several people. In Warumungu, it is the

Box 3.4 Kinship terms in English and in Warumungu (after Simpson 2002)

ENGLISH WORD	WHO IS IT?	WARUMUNGU WORD	WHO IS IT?
Mother	Mother	Karnanti	Mother + mother's sister
Father	Father	Kampaju	Father + father's brothers, father's sisters
Sister	Younger/older sister	Kapurlu	Older sister, + mother's older sister, children of mother's older sister
Brother	Younger/older brother	Paparti	Older brother, + children of father's older brother
Sibling	Sisters and brothers	Kukkaji	Younger sibling
Cousin	Aunt's and uncle's children	Wankili	Mother's brother's children and father's sister's children
Aunt	Mother's and father's sisters		
Uncle	Mother's and father's brothers	Ngarmirni	Mother's brother
Grandmother	Paternal and maternal grandmother	Tapu-tapu	Paternal grandmother, maternal grandfather
		Jurttanti	Maternal grandmother
Grandfather	Paternal and maternal grandfather	Kangkuya	Paternal grandfather

mother's brother, the paternal grandfather and the maternal grandmother that are given distinct names; the other names are shared between individuals. Mother does not have the same individual status in Warumungu, since also her sisters (and her husband's brother's wife) are called karnanti. Similarly, the father in Warumungu is not distinguished individually; kampaju is also the term for his sisters and brothers (and his wife's brother's wife). When it comes to sisters and brothers English has a less differentiated system, since there is no difference between siblings of different ages. Also for cousins, English has a much simpler system. The distinction between different types of cousins in Warumungu is culturally important. There is a primary distinction between cross-cousins (children of your mother's brother, or your father's sister) and parallel cousins (children of your mother's sister or father's brother). Cross-cousins traditionally belong to the group you are allowed to marry, whereas parallel cousins do not qualify as appropriate marriage partners. Warumungu uses three different terms for cousins: (1) cross-cousins are referred to by the term wankili; (2) female parallel cousins are called kapurlu (the same as sister); and (3) male parallel cousins are called paparti (the same as brother). The fact that parallel cousins have the same label as sisters and brothers demonstrates that they are not appropriate marriage partners.

To summarize, in English there is always a gender distinction, whereas the Warumungu system is built upon the concept of marrying group. The system contains more distinctions than English concerning relative age, but fewer when it comes to the gender; many of the terms can refer to both male and female relatives.

Ethnobiology is another research area where cultural influences can be seen. There is a practical need to communicate about the environment and it is much easier if the different animals and plants have names. These names are often based on a combination of different factors, such as how the plant or animal may be used, what it looks like and how it behaves. The usage is described, for example, in old names such as *henbane* (*Hyoscyamus niger*, which has shown to be poisonous to hen). Prominent external features lie behind names such as *seahorse* (*Hippocampus hudsonius*; resembles a horse), *flying fox* (*Pteropus edulis*; a large bat with a face resembling the face of a fox), and *blindworm* (*Anguis fragilis*, a lizard with small eyes, resembling a worm). It is not only the outer looks but also similarities with other animals are taken into account, regardless of whether the different species have a common evolutionary history or not. The way of moving, or other behaviour, is described in names such a *silverfish* (*Lepisma saccharina*, an insect that moves like a fish), and *sting fish* (a fish with poisonous stings). This system, based on utility, looks and behaviour, makes it easy to communicate about plants and animals and to learn and remember terminology.

There is also a scientific method to label and organize the millions of living (and extinct) organisms, the system that Carolus Linnaeus created in the 1700s. His idea was to base the systematization of plants and animals on their genetic relations. The Linnean names consist of several parts; for example family, genus and species. Humans belong to the family *hominidae*, our genus is *Homo* and species is *sapiens* (*Homo sapiens*). This system is used also today, and the Latin names give precise identifications and make it possible for biologists all over the world to communicate with each other.

The ways of classifying are based on different criteria and used for different purposes. The everyday categorization carries important cultural knowledge, useful for others living in the same environment, whereas the scientific classification is used as an international research tool and the cultural information is less clear. The study of ethnobiology is important not only to gain an understanding of human cognition, but also as important linkage to the accumulated human knowledge. Since names of plants may reflect their earlier use, there is a risk that specific knowledge gets lost when the words and the language disappear.

3.3.4 But what about grammar?

After this excursion into words and worldviews we will come back to the language-learning child and the acquisition of grammar. One of the most prominent child language researchers, the American psychologist Roger Brown (1973) went out to find children who had just started with multi-word utterances and recorded them on a regular basis until they had reached a stage where multi-word utterances were common. The research is based on spontaneous speech data, taken from a longitudinal corpus following language development of three children, Adam, Eve and Sarah. These children were recorded in interaction with their parents, Adam and Eve every second week, and Sarah each week. Eve was 18 months, and Adam and Sarah were 27 months when the project started. Thus, the children differed in age of onset of multi-word utterances, but, importantly, they were similar in the subsequent development. Based on the results of the analyses from these three children (and also additional data), Brown (1973) suggested a stage model for the development of sentence structure, a model which is still in use, for example in the assessment of language disorders.

In his book *A First Language* (1973) Brown discusses data from children speaking different languages. The framework he uses is based on semantic relations. Semantic relations are suggested to be what keeps early multiword utterances together, such as agent-action (*mummy read*), possessor-possession (*mummy sock*). Box 3.5 illustrates the most frequent semantic relations found in the two-word utterances of English-speaking children. These are used also in children speaking other Indo-European languages such as Swedish, and Mexican Spanish, but also in totally unrelated languages such as Finnish (a Finno-Ugric language) and Samoan (Austonesian). Such cross-linguistic similarities would support a universal first stage in an emerging grammar, a stage that is independent of specific features of their particular languages, and it could have been a candidate for an innate basis for grammar in human, if it had not been for the fact that the same semantic relations accounted for most of the examples (78%) of the data from the language-taught chimpanzee Washoe (cf. Chapter 5.8.2) This suggests that it is rather a matter of what caretakers are talking about, which makes up the frequent patterns in the input to the child (and the chimpanzee). The box below illustrates the eight most frequent semantic relations.

Box 3.5 Semantic two-term relations in early speech (after Brown 1973)

SEMANTIC RELATION	EXAMPLE	SEMANTIC RELATION	EXAMPLE
Agent – Action	Mummy read	Entity – locative	Book table
Action – Object	Read book	Possessor – possession	Mummy sock
Agent – Object	Mummy book	Entity – attributive	Big dog
Action – Locative	Sit here	Demonstrative – entity	There ball

Observe that the words do not have to come in the order given above. *There ball* may as well be *Ball there*, and *Big dog* may as well be *Dog big*.

In the analyses of semantic roles above, it is striking that so many grammatical markers are missing. The children say *read book* instead of *read the book*, *book table* instead of *the book is on the table*, *mummy sock,* instead of *mummy's sock*. Here, we will account for how these grammatical markers come into the child's speech.

The emergence of grammar is related to vocabulary size – as the vocabulary grows, the grammatical markings become more common. Not only does the child get more words, but the words are also of different types. The earliest child vocabulary is dominated by content words from the open classes, such as nouns (*Mummy, car, dog*) and verbs (*fall, look, gone*), produced in short utterances, often one- or two-word utterances. Later, closed-class words emerge, such as adverbs (*here, there, now, maybe*), prepositions (*in, on*), pronouns (*me, she*), and conjunctions (*and, while*). These words are also called grammatical words, since they carry little meaning but express grammatical functions. When grammatical words are used, the child's production changes from "telegraphic speech", to grammatical utterances. Before the "window is opened to grammar", the child is not able to reuse the constructions used by adults according to the alignment processes mentioned above.

The two examples from Carl below (taken from the CHILDES database, the Manchester corpus), will illustrate how child-adult interaction at the same time resembles and differs from adult-adult interaction. The alignment between the speaker's utterances is there, but Carl makes systematic changes to the mother's constructions. In Example (1), Carl is 1 year and 8 months old (1;8), and his grammar is not developed enough for him to use some of the grammatical structures that his mother is using. For example, Carl repeats 'what're they doing', but without using the copula verb *(a)re* (instead he says 'what they doing').

Example 1. Carl 1;8

*MOT:	what're they doing Carl?
*CARL:	what they doing?
*CARL:	what they doing?
*MOT:	no what are the cars doing?
*CARL:	what you doing.
*CARL:	there the bus.
*CARL:	there the door.
*MOT:	they're going through a door?
*CARL:	that way.
*CARL:	that way.
*MOT:	let Mummy get them.
*MOT:	they're coming back.
*CARL:	coming back.

After 8 months, Carl's grammar is more developed and he now uses the copula verb consistently. In Example (2) he uses the copula in all places where it could be expected (Mummy's coming, he's gone, it's Nana).

Example 2. Carl 2;4

*MOT:	that's their Mummy coming.
*CHI:	Mummy's coming.
*MOT:	and Bouncer's running away.
*MOT:	oh dear.
*MOT:	look.
*MOT:	he knocked Grandpa over.
*CARL:	down the
*CARL:	**he's gone.**
*MOT:	what's that?
*CARL:	Daddy.
*MOT:	oh you can hear Daddy.
*MOT:	no.
*MOT:	that's Granny.
*MOT:	who's that?
*MOT:	is that a lady?
*CARL:	no.
*CARL:	**it's, it's Nana.**
*MOT:	**it's Nana?**
*CARL:	mmhm.

The examples show instances of reusing words from the interlocutor, both from Carl and from his mother. In Example (2) (Carl 2;4), Carl is reusing 'Mummy coming' and his mother is reusing 'it's Nana'. Language acquisition is a process where children have to construct their own versions of the target, and they do so by gradually adjusting to the environment. In the beginning there are systematic differences due to the child's constrained grammar, but as the child's language ability develops the differences get smaller and smaller.

3.3.5 Recursion

Recursion means "the capacity to generate an infinite range of expressions from a finite set of elements" (Hauser et al. 2002:1569) and it has been proposed to be the "only uniquely human component of the faculty of language" (ibid.) How and when do children start using this?

In comparison to other language skills, recursion is a rather late phenomenon, and understanding of recursive patters develops slowly. Even at the age of between 3 and 4 years, when children already use subordinate clauses, they find it difficult to understand the combination of adjectives such as in the phrase "Show me *the second green ball*". When adults are put to this task they take one of the balls, (namely the second green one), whereas children tend to interpret it as a conjoined phrase (the second *and* the green) and take two balls, the second one and the green one (Roeper 2011). Roeper suggests that there are several phases in the acquisition of full recursion; first single adjectives (*the second ball*) then direct-conjoined recursion (*the second ball and the green ball*) and finally indirect recursion (*the second green ball*).

On the sentence level, recursion is used for example when we talk about what others' say and think. When reporting what others say it is common to use the complementizer *that*, as in *They said that she had a new car, He said that they said that she had a new car, I said that he said that they said that she had a new car*. This can continue as long as it is possible to remember who did what – there is no grammatical rule putting an end to how many clauses can be added to the list.

Talking about what others say starts early in the interaction with children. A favorite game is to talk about what animals say: *what does the dog say, the cat, the duck, the car, the clock*, etc. Another example of recursion occurring in the lives of young children is nursery rhymes of the type "The house that Jack built". This is the kind of cumulative tale that is often told to, and together with, pre-school children, who seem to enjoy remembering the different parts. The rhyme is built on the repetition of embedded clauses (of the relative clause type) stringed together. It would not have been possible without the complementizer *that*, and each clause that starts with *that* is an embedded clause. The recursion starts already in the first sentence "This is the house that Jack built", which contains the noun phrase "the house that Jack built", which refers to a particular house. And then the story goes on and on.

> This is the house that Jack built
> This is the malt that lay in the house that Jack built
> This is the rat that ate the malt that lay in the house that Jack built
> This is the cat that killed the rat that ate the malt that lay in the house that Jack built
> This is the dog that worried the cat that killed the rat that ate the malt that lay in the house that Jack built

Like when we report what different people say, we can add on more and more clauses, until we are unable to remember more (in fact, the story of Jack above has more verses than are shown here).

A phenomenon that is related to recursion is the concept Theory of Mind, which refers to an understanding of the minds of others. The verbal expression showing that someone has Theory of Mind is usually the use of embedded clauses *I think that she thinks …* It is generally agreed that a Theory of Mind, can be found in human children from around 4 years of age, but there is no consensus about its presence in other species (for an overview of primate research, see Call & Tomasello 2008). To test whether young children have access to a Theory of Mind something like to the following scenario is often set up: "Maxi puts a chocolate in a cupboard. Then Maxi leaves the room and his mother enters. She takes a piece of the chocolate and puts it back – but in another cupboard. Then Maxi re-enters." The question for the child to answer is: *Where will Maxi look for the chocolate?* The child is asked to point at the place where Maxi will look (Wimmer & Perner 1983). Younger children typically point to the place where the mother put the chocolate, thus showing that they are not aware of the fact that people with different experience can have different information.

Theory of mind is not only related to both social experiences and cooperation, but also directly linked to language and language use. Some facets of theory of mind are directly connected to syntax. For example, utterances like *I know that he knows, you know that I know that he knows,* which are often used in everyday speech, illustrate syntactic recursion, and it is hard to imagine how that content would be expressed without language.

3.4 When problems arise – Specific Language Impairment

In the previous, language acquisition has been described as the most natural process in children, something that comes easy. However, there are also children with problems in acquiring their mother tongue. According to the SLI Consortium (2002), approximately 4% of English-speaking children are diagnosed as having Specific Language Impairment (SLI), "a disorder in the development of language skills despite adequate opportunity and normal intelligence" (SLI Consortium 2002: 384). The language disorder is manifested for example in grammatical problems that are more similar to second language acquisition (Håkansson & Nettelbladt 1996, Paradis & Crago 2000). Some of the problems seem to be linked to processing problems and a shortage of phonological memory storage, since one of the best indicators of SLI in children is that they find it difficult to imitate nonsense words (Gathercole 2006, Leonard 1998).

The first studies of SLI had practical aims and pointed towards contributions to clinical methodology, to help the children acquire their language. The underlying causes of SLI were often thought of as being a result of the environment, i.e. a problematic family situation, or ear diseases in early infancy. However, new techniques for genetic analyses made it possible to try out what many had suspected – that SLI runs in the family – and the last decades we have seen an increasing interest in the possible relationship between genes and language. This made the SLI question interesting not only for people working with language acquisition, but also highly relevant in the discussion of language evolution, and what makes us humans. One way to find out is whether there is a genetic factor in SLI is to study twins. Bishop and colleagues (Bishop et al. 1995) compared monozygotic twins (from one egg) to dizygotic twins (from two eggs) and found significantly more monozygotic than dizygotic twins with SLI that had a sibling with SLI. The greater similarity between monozygotic twins than between the dizygotic twins suggests a genetic influence. Sometimes only one of the twins was diagnosed with SLI, whereas the other had other communicative problems, not met by the criteria for SLI. This means that there may be several factors involved – only if there is more than one risk factor at hand, the child would be identified clinically as SLI. Bishop (2006: 220) suggests that there may be more than one route to effective language acquisition. If only one route is blocked, there may be an alternative way to use, but if several routes are blocked, language problems will show up.

Two important break-throughs emerged in the search for genetic explanations; first the finding of a family, KE, where half of the members were affected with severe language problems (Hurst et al. 1990, Gopnik 1990), and then the discovery that a particular gene, the FOXP2 gene was disrupted in all the affected individuals (Lai et al. 2001). This gene is involved in embryonic brain development, and it has been found to be active also in other species, for example in nonhuman primates, mice, and songbirds (e.g. Haesler et al. 2004). In humans, however, it is assumed that some functional changes in the expression of this gene took place already in the Neandertals (Krause et al. 2007), which makes it particularly interesting for the issue of language development and impairment. The KE family consists of over thirty members, from three generations. Another patient, unrelated to the KE family, has been found to exhibit the same pattern, language impairment and disruption of FOXP2. On the basis of these findings, it has been concluded that FOXP2 plays an important role in language development, and that if it is disrupted in early stages this "leads to abnormal development of neural structures that are important for speech and language" (Lai et al. 2001: 522). In Zeesman et al. (2006) another patient is discussed, a 5-year-old girl with similar language problems, plus problems in coughing, sneezing and laughing. Also in this case a genetic explanation is given, this time in an area neighbouring to FOXP2.

However, while many scientists have gone further in defining the exact functions and structures of FOXP2 (e.g. Vernes et al. 2006), there are also findings that indicate that the relationship is not as straightforward as was assumed at a first glance. For example, the SLI Consortium (2002) examined FOXP2 in individuals from 98 families with known language disorders, and found no mutation of this gene. What they found was a linkage to another chromosomal region. Also Newbury and colleagues (Newbury et al. 2002) reported findings that the FOXP2 is not underlying language disorders in general.

Concluding the discussion on language and genes, it is important to keep in mind that language capacity is a multifaceted phenomenon and that there are probably many factors involved in language development as well as in language impairment. Even though the finding of the FOXP2 gene does not tell us everything about language, it has given many new insights. Fisher (2005) points out that although FOXP2 does play a role for speech and language, it was important already in the shared ancestor of humans and rodents, and it "cannot be characterized as the 'gene for speech' but rather as one critical piece in a complex puzzle" (Fisher 2005: 111). Lieberman confirms this view and adds a note of hopeful anticipation: ". (..) other regulatory genes undoubtedly are involved in the evolution of human language and cognition, but the gene provides an opening into our understanding of the nature of the neural bases and time course of the evolution of these human qualities." (Lieberman 2006: 126).

3.5 Second language acquisition – organizing language once again

What happens then when humans learn another language later in life? As mentioned before, the human language faculty involves a potential to acquire more than one language. There is in fact no limit as to how many languages a person can acquire. Some learn two languages simultaneously as first languages, whereas others first acquire one language, and later additional languages, up to ten-fifteen languages. There are some intriguing differences between first and second language acquisition, the most important being that in first language learning, children learn how to mean and how to speak at the same time. Joseph (2004: 184) suggests that first language users use language both for thinking and for communicating, whereas second language users use language only for communication. We acquire our world-view and identity by our mother tongue, and the culture we are born into, but the second, third and fourth languages are learned in order to communicate with speakers of those languages, not in order to help us think. Second language learners already have their identities formed and their task is to find new ways to achieve what they are already able to do in their first language.

Similarly to the modifications that have been found in adults' speech to infants (*Baby Talk*) speakers modify their speech to adult language learners in systematic ways. *Foreigner Talk* is the term for the speech modifications made by native speakers addressing non-native speakers with low command of the target language. The native speakers usually modify their pronunciation by a slower speech rate and clearer articulation. They use a less varied vocabulary, with many so-called international words (e.g. *problem, finito, amigo, kaputt*) they simplify their grammar and the whole interaction is different by having a lot of repetitions and comprehension checks. Language teachers tend to use *Teacher Talk* when addressing beginner learners. The speech tempo, sentence length, and number of subordinate clauses increase steadily during a language course. The correlation found between amount of modifications and comprehensibility of the narrative by the non-native listeners, shows that modifications increase intelligibility (Håkansson 1987).

The development of a second language is both similar to and different from the development of a first (Meisel 2011). Lexical acquisition is similar, as also adults may use fast mapping (Markson & Bloom 2001). Acquisition of phonology and grammar show differences. There is often a strong influence from the first language in pronunciation, as learners preserve the sound patterns from their first language, which is recognized as an foreign accent. However, this is not always the case. Some learners manage to adapt to the phonology of the second language quite well, even if they learn the language as adults. The earlier the second language is acquired, the higher is the chance (or risk) that it will in fact replace the first language. As mentioned earlier (Chapter 3.2.2) the sounds of the first language may get totally lost, and replaced by the speech sounds of the second language in internationally adopted children. The processing of a second language in the brain has also been studied using Event-Related Potentials (ERP). Steinhauer et al. (2009) demonstrated that advanced second language learners, even if learning the target language after puberty, processed language in the same way as native speakers. The beginning learners, on the other hand, showed different brain activities. Intermediate learners differed again, showing delayed reactions, but of the same kind as native speakers. This suggests that it is possible to process that second language like a native speaker, measured by ERP, at an advanced stage of learning – which is against the idea of a sensitive period for second language acquisition.

Processability Theory (Pienemann 1998, 2005) assumes that the second language learner has to acquire procedural knowledge for the constructing of utterances in the second language. The theory builds on Levelt's (1989) model of speech processing in native speakers. In order to be able to plan and perform an utterance at the same time as taking in what others say, the speaker must have access to automatic grammatical procedures. These procedures are created in an incremental order, so that one procedure may serve as the input for next, from words, to phrases, to clauses. The procedures are used in an automatic way by native speakers, but they will have to be recreated and organized in the process of acquiring the second language. Before the learner can process the grammar on a particular level, it is not possible to align to the native speaker. For instance, a learner who has not yet reached the level of third person –s (Level 4 according to Processability Theory) uses either infinitive (*he help*) or progessive form (*helping*) of the verb *help*.

Like in other dialogues, second language learners participating in dialogues align with the interlocutors. But what happens if the beginning second language learner does not share the same linguistic knowledge as the interlocutor? The examples below are taken from the ESF corpus (lpera16i.ltr.cha from CHILDES database), with the learner Ravinder. The dialogues show that Ravinder has not yet reached level 4 (3 person –s).

Example (3) below illustrates successful alignments, where the learner is able to pick up and reuse phrases like "old home", "the end" and "happy". These phrases and words do not present grammatical challenges for the learner.

Example 3. INT: native speaker interviewer, RAV: learner "Ravinder"

*RAV: charlie and girl and with home hes all broke thats it is not too very good xxx.
*INT: i see its an **old home**.
*RAV: oh yeah **old home**.
*INT: mhm mhm.
*RAV: and hes live together.
*INT: mh.
*RAV: and dinner.
*INT: mhm.
*RAV: thats it and er dinner is finish or walk on the road.
*INT: mm.
*RAV: m that finish.
*INT: thats **the end**?
*RAV: yeah.
*INT: uhuh.
*RAV: **the end** yeah.
*INT: so it was a **happy** ending?
*RAV: yeah **happy** very happy.

However, Example (4) shows what happens when the linguistic structures are out of reach for the learner. The native speaker does not align to the utterance "charlie help" but reformulates it into "charlie helps her". Despite the fact that the learner is given this form, he does not use it but reformulates it into the progressive form "helping" instead. The progressive form belong to stage 2 in Processability Theory. Later in the example, the learner produces "hes not pinch erm, me pinch charlie say". Here, it is obvious that the native speaker does not align to the form, but he aligns to the content. The utterance is reformulated into "he said: she did not pinch it, I pinched it". Just as adults do when talking to children, it is common for native speakers to reformulate ungrammatical and uncomplete utterances and make them grammatical and targetlike.

Example 4. INT: native speaker interviewer, RAV: learner "Ravinder"

*RAV: erm she not work.
*INT: mhm.
*RAV: hes pinch erm from bread.
*INT: mhm.
*RAV: **charlie help.**
*INT: **charlie helps her?**

*RAV: yeah.
*INT: mhm.
*RAV: **helping girl.**
*INT: mhm.
*RAV: **hes not pinch erm me pinch charlie say.**
*INT: oh i see **he said she didnt pinch it i pinched it.**

The last example shows that the learner is not able to align with the native speaker, and it also shows that the native speaker is not aligning to the learner. This is common in cases when learners produce ungrammatical utterances (Costa, Pickering & Sorace 2008), and can be expected to help the learner to realize that there were some problems.

Second language learning has been suggested to take place outside of a critical period. Instead, second language learners "can be seen not as the unfolding of some prearranged plan, but rather as their adapting to a changing context, in which their language resources themselves are transformed through use" (Ellis & Larsen-Freeman 2006: 578). As we will see, this view of second language development bears similarities to some of the descriptions of bird song acquisition, for example in the hierarchical learning of the complex song of the nightingale (Hultsch & Todt 2004, Todt 2004, cf. Chapter 7), and probably it is this type of language acquisition that should be discussed in experiments with language teaching to other species.

3.6 Sign language – another modality

Until now we have focussed on spoken language, but there are also other possibilities to use human language. Deaf and hearing-impaired people use other means of communication than the vocal-auditive. There are different types of signing systems, for example the "hand talk", used among American Indians, finger spelling, where each sign represents a letter of the alphabet in a spoken language, and sign languages, which constitute languages in their own right. In this section we will describe sign languages. Sign languages have referential and expressive functions, lexical forms and grammar rules, just like spoken languages, but the difference is that they are gestural-visual instead of vocal-auditive. There are many misconceptions about sign language, for example that they would be universal, and that they are versions of the spoken language. Both these ideas are false. Sign languages are not universal, and they do not correspond to the national spoken language. There are many different sign languages, for example American Sign Language, Australian Sign Language, British Sign Language, Chinese Sign Language, Danish Sign Language, Finnish Sign Language etc. American Sign Language and British Sign Language are not mutually comprehensible. ASL (American Sign Language) grammar differs from the grammar in spoken American English for example in the use of pronouns and in the use of classifiers (Liddell 2003). Typically, pronouns are indexical in sign languages, which means that they point to the referent, or the place of the referent. The ASL pronominal system distinguishes between singular, dual, multiple and plural, whereas in spoken English there is only singular and plural. Classifiers are used in many spoken languages (but not English) and they are also common in sign languages. One common function is to give information about size, shape and texture of an object by classifying objects into groups.

Signs are formed by handshapes, movements and locations and they can also be combined with facial expressions. Sign languages have words and word categories, just like spoken language, and the words are made up from smaller parts. In speech, the words are based on speech sounds, whereas in sign language the words are made up from a combination of handshapes, locations and movements. They can also be combined with facial expressions. The form of a sign can sometimes be iconic, i.e. resemble the referent.

The acquisition of sign language is similar to the acquisition of spoken language in terms of timing and developmental milestones (Baker, van der Bogarde & Woll 2008). The child constructs the language from the input given in the environment. As mentioned above, deaf children vocalize like hearing children during the very first months after birth, but the vocalizations decrease and they do not reach the stage of syllabic babbling. Instead, they use gestures in a way that has been analyzed as "manual babbling" before they start using signs for words. All children move their hands and arms, but deaf children that have been exposed to sign language use a larger variation of handshape types and movements (Petitto & Marentette 1991). The first meaningful signs emerge around the age of one year. Similarly to hearing children who have a "childish" pronunciation before they have the motoric skills, young deaf children may use gestures diffently from the adult use, for example by using the whole hand instead of only the index finger in a gesture. The early vocabulary is the same as for children learning spoken language, i.e. words about things that are important in the surrounding environment, family members, food, animals, toys. The first two-word combinations are usually pointing together with a hand sign. After this period, combinations with several signs are produced. Some phenomena are acquired later in the development, for example the use of classifiers and the facial expressions combined with the manual gestures (Mayberry & Squires 2006).

3.7 Summary

There are 6000–7000 different languages in the world. One reason for this large number is the fact that identity is often expressed through language. To feel group solidarity it is important to share linguistic code. This is particularly clear in multilingual societies. Language functions are similar across the world, but when it comes to formal properties of language fundamental differences appear, showing the potential for variation. Studies of semantic fields demonstrate that there are cross-linguistic differences in how for example kinship relations are conceptualized. In this chapter we have demonstrated some of the many ways to express different meanings and illustrated with languages from various parts of the world.

The acquisition of language in the context of first and second language acquisition has been discussed, and the phenomenon of language impairment and its possible relation to genetic factors has been presented, as a potential example of a genetic basis of human language. Using language is an undertaking of cooperation, where speakers align to each other. Children learn language by gradually constructing the system, in close cooperation with adults. The developing grammar grows gradually and it is, just like phonological development, characterized by stages. Around the age of three–four years, children seem to have grasped the basics of their grammar. As an explanation for the smooth development it has been suggested (by Chomsky and colleagues) that children are equipped with an inner model of the language, a Universal Grammar, which they match the language of the environment against. This idea of children matching input to an inner model bears some resemblance to what young songbirds do (see Chapter 7). We will return to this discussion in Chapter 8.

3.8 Suggested readings

Evans, N. (2010). Dying words. Endangered languages and what they have to tell us. Malden, MA.: Wiley-Blackwell.
In Dying words, Evans gives an overview of phenomena in the languages of the world – with particular reference to languages that are on the verge of disappearing. He makes an appeal to document these languages before it is too late, to save the knowledge of human life that they contain.

Crystal, D. (2006). How Language Works: How Babies Babble, Words Change Meaning and Languages Live or Die. London: Penguin.
David Crystal manages to get the essence of linguistic study together in one volume: language development and use, structure and change, writing, signing and speaking. The combination of scientific details and anecdotes makes it a both informative and appealing introduction to linguistics.

Pinker, S. (1994). The Language Instinct. How the Mind Creates Language. New York: Harper.
This book discusses language as a human instinct "just like the spinning of a web is an instinct in spiders". It is written in an engaging style, provocative as well as entertaining, and accessible also for readers with no background in linguistics.

Chapter 4

Human nonverbal communication

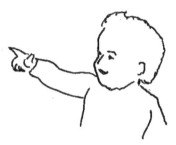

4.1 Introduction

Language is not the only way to communicate for humans. Nonverbal communication, i.e. communication without words, is a significant means to share information. The cooperative side of communication is in fact more prominent in nonverbal than in verbal expressions, and it is at its best in the synchronization of behaviour. Nonverbal expressions play an important role also during an ordinary conversation. At the same time as the speaker is articulating words and sentences, information is also conveyed by gaze, tone of voice, facial expressions, gestures and postures. This is done in such a smooth and efficient way that we are hardly aware of it. Moreover, it seems to be impossible not to send and receive that information.

The social life of humans (as for other group-living animals) puts large demands on communication, and one function is to coordinate activities of the group. There are expressions for communicating group solidarity as well as expressions for showing individual identity and position in the hierarchical organization of the group. Co-operation is crucial in the rearing of the young, in food searching, and protection. Individual identity is imperative to recognize family, to position oneself in relation to the group, to avoid aggressive confrontations. In situations when solidarity is needed, people converge towards others in the group – for example, gestures and dress codes are often used to signal solidarity. In the opposite situation, people show dominance by upright body posture, or non-dominance by stooping head and shoulders. These expressions are not static. They occur as parts in a dynamic interaction, and each attempt to isolate entities results in loss of the interactive dimension that is the basic element in all communication. Just as people make interactional alignments in dialogues, there is a constant flow of mutual adjustments to the others in the non-verbal dimension. Signals from A are met by signals from B and C, which may affect the behaviour of D, which in turn may have an influence on A. Power relations are constantly negotiated in this manner. There are two linguistic perspectives on synchronizing behaviours: within sociolinguistics it is common to talk about speech accommodation as expressions of group solidarity, whereas psycholinguists talk about alignment in their descriptions of online speech production. Synchronizing of behaviour is an important issue in research on nonverbal communication, and it is demonstrated by body postures, hand gestures, and facial expressions.

4.1.1 Research methods and observation techniques

Unlike the models for human verbal languages, where linguists agree on units like sounds, words, phrases, and clauses, there is no consensus as to how nonverbal expressions can be systematically described, or which units to use. The nonverbal system does not contain the same features as verbal language (with discrete signs where each sign has content and form) but the expressions are often continuous (i.e. degrees of anger can be expressed by increasing the same expression). There is a need of another analytical framework. It may seem reasonable to start from the topology of the body, like Darwin did in his book *The expression of the emotions in animals and men* (1872/1965). This is the way animal expressions are often categorised, for example with focus on if it is the tail or the face that is expressing something. Another categorization principle is to use the sensory channels as a starting point. Then the expressions are classified according to the sense that is receiving the signal (e.g. acoustic, visual, tactile and chemical/olfactory expressions). A third way is to base the classification on the origin and function of the behaviour. This is what is proposed by the America psychologists Paul Ekman and Wallace V. Friesen (1969). They identify five different categories that can be placed along a continuum from culture-specific to universal. A fourth method is to base the units of classification on movement patterns. The British scholar Adam Kendon is one of the leading specialists on gestures and he suggests the following definition for gestures that are used simultaneously with speech: "This entire excursion, from the moment the articulators begin to depart from a position of relaxation until the moment when they finally return to one, will be referred to as a gesture unit." (Kendon 2004:111). This classification method makes it possible to streamline speech and gesture in a given speech situation.

Before the time of instrumental recordings, expressions of human nonverbal communication were noted and written down. This has the disadvantage that some behaviours will not be noticed. A more objective method came with the possibility of audio or video recordings. Filming makes it possible to perform analyses frame-by-frame, and study expressions over and over again and in great detail. Facial expressions can be measured by the Facial Action Coding System (FACS), which was developed by Ekman and Friesen (1978). The system is based on early work by Duchenne (1862/1990), and links muscle movements to facial expressions by first stimulating the muscles by electrodes and then photographing the face. The FACS system has been successfully adapted to facial expressions of chimpanzees, under the label ChimpFACS (Vick et al. 2007), which made it possible to compare muscle activities in humans and chimpanzees.

Another method is to investigate how nonverbal expressions are decoded, or interpreted. This is done by asking subjects to describe how they perceive nonverbal expressions. Vocal and facial expressions, body postures and movements are presented to the informants and their task is to describe what they are seeing (Is this person in the picture happy, sad, etc?). For example, informants look at pictures of faces to decide their emotions and attitudes, or informants watch video clips of others communicating and rate aspects of the interaction.

Recent developments in neuroscience have paved the way to investigate which areas in the brain are active in responding to different stimuli. By use of fMRI (functional Magnetic Resonance Imaging) it has become possible to study responses to nonverbal expressions in the brain, for example to compare visual to acoustic stimuli (Beneventi et al. 2007). Similarly, methods such as EEG (electroencephalography) and ERP (Event-Related Potentials) may be used to indicate when the brain reacts to gestures.

Finally, talking about coding systems, it is important to keep in mind that the division between verbal and nonverbal communication is an artificial division, made by the analyst – in real communicative situations gestures and language cannot be separated but are combined into multimodal expressions.

4.2 Functions – what do we use nonverbal communication for?

Traditionally, nonverbal communication has been interpreted as expressions of emotions and interpersonal attitudes. The interactive dimension is important – emotions and attitudes are not expressed in a vacuum but are socially anchored. Many expressions have as their function to form and keep social bonds, for example, by smiling and touching, and also to show interactional synchrony by mirroring postures, gestures and facial expressions of others. Some expressions are culturally determined. The cultural filter is particularly conspicuous when it comes to ceremonies such as greeting and parting behaviours. These behaviours differ between cultures. They serve at the same time as identification markers and as releasers of ritual patterns in other participants of the same sub-culture. Another situation where nonverbal communication plays a key role is in the courtship behaviour. To have a well functioning courtship behaviour is of vital importance for the survival of the species. One of the distinctive features in human courtship is the intense eye contact, interrupted with occasional sideways glances. Studies comparing different cultures have demonstrated that the invitational postures are similar between cultures but differ between females and males; females typically present the breasts by drawing back or raising the arms, whereas males invite by squaring the shoulders. Rejecting behaviours include pulling back, crossing the arms, frowning and avoiding eye contact.

However, nonverbal expressions are not restricted to the affective and social functions. The repertoire of nonverbal expressions also contains expressions that function referentially, directing the attention of others' to some object or event outside the individual, for example by pointing or by illustrating something by mimicking. The study of pointing in different cultures has demonstrated distinct patterns (Kita 2003). Pointing can be performed together with, or independent of speech and carried out by using hand, finger, head, lips, or by gaze. The object of interest may be an actual object in the environment, which can be found if one follows the direction of the hand, but it may also be an invented and/or invisible object. For example, a speaker may point at an invisible map to show the relative location of characters in a story, or at the sky to indicate where the sun will be at a certain time. Indexical gestures like pointing are often accompanied by some kind of eye contact, typically looking at the indicated entity and back to the interlocutor (more about pointing and gaze below).

4.2.1 Permanent versus temporary expressions

In animals, some expressions such as body size and plumage colour are considered permanent signals. In humans it is a complicated issue to decide if expressions are permanent or temporary, since people tend to consciously manipulate their bodies in various ways to change size and structure of body. A large body is associated with strength and therefore generally also with dominance. However, there is no one-to-one relationship between height and leadership among humans; i.e. not all leaders are tall. Studies on leadership among adolescents have demonstrated that there is a certain tendency for leaders to be taller than followers, but there is not an absolute difference; some leaders are even shorter than their followers (Granström 1992). We know this also from history – for example Napoleon is said to have been short.

One way to change physical appearance is by exercise; another is by the way of dressing. One may for example add on to one's height by having extra inches on the shoes or wearing high hats. One of the key signals of male presentation is shoulder width. To increase shoulders it is common for people to wear feathers, shoulder pads and epaulettes.

> In men broad shoulders are desirable, and rarely will we find a hero in art or literature who has narrow shoulders. The width of the shoulders in relation to the narrow hips is very effective, although it may be tremendously overdone. (Eibl-Eibesfeldt 1970: 434)

4.2.2 Synchronising in sympathy

As mentioned earlier, accommodating and aligning to others by synchronising speech rate and other features is frequently observed in human verbal interaction (for example van Baaren et al. 2003, Pickering & Garrod 2006). The same phenomenon occurs in gestures and postures, and is known under various labels such as mimicking, mirroring, imitation, behaviour matching, and the chameleon effect. These terms have been used to indicate essentially the same meanings, one difference being that the term synchrony is usually related to temporal and rhythmic timing, whereas the other labels often refer to the similarity in how the gestures are performed. The discussion of synchronization took a new turn when Rizzolatti and Arbib (1998) discovered so called mirror neurons in the brain, neurons that simulate a copy in the brain of an observed action, or action-related sound, and can be presumed to lie behind synchronization (Bråten 2002, Rotondo & Boker 2002, Hauk et al. 2004).

Children start synchronizing their behaviour very early, and synchronization of certain behaviours, such as tongue and lip movements have been found to occur as early as at the age of three days (Meltzoff 1981, Meltzoff & Moore 1977). Synchronizing behaviour is typically found among individuals belonging to the same group, and most studies have focused on mirroring in couples, groups of students, and friends. Close friends can synchronize movements and gestures, automatically and with remarkable precision. However it is also possible to experimentally elicit mirroring in people by suggesting the prospect of a relationship. Lakin and colleagues (Lakin et al. 2003) created a situation where informants were given affiliation as the goal of the interaction; for example, they were told that they needed to cooperate with another person to fulfil a task. Situations like that trigger imitation, and also situations where personal matters and relationships are discussed. Mirroring each other underscore the solidarity between the interactants. It is more conspicuous when talking about positive than negative events, possibly because positive emotions are acted out in a more powerful way. In an experiment where students were asked to tell each other about the happiest or the saddest events, the results demonstrated synchronization both in the positive and the negative contexts, but stronger in the positive settings (Kimura & Daibo 2006).

Synchronization of rhythmic behaviour has been found in different kinds of body postures or movements, such as the way of standing, sitting, walking; movements of the hands, the feet, the lips (e.g. Kendon 1975, Dimberg et al. 2000). It is also shown in laughing, weeping and yawning. From a social perspective, mirroring can be described as forming the glue of solidarity in the group. This can potentially have had a strong evolutionary significance. There is a strong survival value attached to the maintenance of group relationships, with the benefits of belonging a group, such as protection, and sharing of resources.

4.3 Acoustic communication

The importance of human acoustic communication is indisputable – infants recognize speech sounds and start to vocalize immediately after birth. This is a whole year before they start using words. The voice has more qualities than bearer of words, and provides a lot of information about the speaker, such as gender, age, emotions, and attitudes. It gives information both independently from speech so-called *extralinguistic* information (for example laughter, cries, coughing), and together with speech, so-called *paralinguistic* information (for example tone of voice). In this section, we will first introduce the extralinguistic type and then go over to the paralinguistic kind.

4.3.1 Extralinguistic expressions – laughters and cries

Acoustic expressions are used in human communication also outside the context of speech. The human laughter is probably an innate expression, since also deaf-blind children produce laughs without having had the possibility to learn from exposure, i.e. neither hear nor see anybody laugh. Laughter is not only used in humorous situations, but has a clear social function and is used to release tension, and to show submission (e.g. Vettin & Todt 2004). Acoustic analyses of laughs show that there are different types, for example voiced and non-voiced laughs. The voiced laughs sound more melodic and song-like, whereas non-voiced laughs give a noisy impression and are produced with the mouth wide open. Females tend to laugh more in interaction with males than with other females (Owren & Bachorowski 2003), and antiphonal laughs (i.e. laughs that are synchronized with, or occur immediately after somebody else's laugh) are used significantly more often among females than among males (Smoski & Bachorowski 2003)

Crying with tears is assumed to be the only emotional expression that is uniquely human. A crying face brings out strong reactions from others, who respond by trying to stop the crying by various means. In this way, crying functions as an immediate attachment bond, both in children and adults (Hendriks & Vingerhoets 2006, Nelson 2005).

4.3.2 Paralinguistic expressions – with speech

The voice is a unique characteristic of an individual and we recognize people by their voices, just like we do by looking at their faces. This is evident even in very young children. Newborn infants are found to recognize their mother's voice two hours after birth (Querleu et al. 1984). In fact, is so easy to be recognized by the way of speaking that criminals often try to hide their identity by not talking, or by trying to change the voice in different ways. The voice conveys information about personal features such as gender, age, and geographical origin. Gender is maybe the most apparent indication, due to anatomical differences in the vocal tract. Men talk louder, with lower pitch and less pitch variation than women. Dialectal features play a fundamental role, and dialectal features take priority over many other properties of an individual's voice in the identification (Hollien 2002, Zetterholm 2003). Another variable that is discussed in the literature is age. With increasing age, the speaker uses lower pitch and slower speech rate, the voice gets harsher and more strained (Linville 2001, Schötz 2006). Age is assessed by different cues in female and male voices. In females prosodic features are more important, whereas spectral features seems to be more important for male age. Around puberty, the vocal tract length starts to increase in males and continues until the individual reaches full maturity, lowering the fundamental frequency. Studies have shown that females prefer male voices with lower fundamental frequencies, which signals masculinity. Feinberg et al. (2005) manipulated fundamental frequencies in male voices, and found that females preferred the voices with lowered fundamental frequencies to the voices with raised fundamental frequencies.

Together with a semantic content, the voice may express emotions. Listeners are able to recognize emotions on the basis of voice alone, independent of the verbal content. Generally, higher pitch is associated with happiness and surprise and lower pitch with disgust and boredom. Children do not learn to reliably interpret emotional expressions of others until around the age of 4, possibly due to a late development of theory-of-mind, i.e. the ability to understand states of mind in others. Some affect displays, for example anger, seem to be interpreted earlier than others, however. This is associated with the crucial difference between a threat and a submissive fear-expression, typically expressed by the so-called "frequency code", shared among mammals and birds (Morton 1977, Ohala 1980). There is probably a selective value in the capacity of feigning different body sizes to deal with different situations in social life. Interestingly, experiments have shown that humans intuitively match sound to size when exposed to unfamiliar sounds (Coward & Stevens 2004).

The voice of the speaker also reveals whom the speaker is addressing. For example, speakers increase the fundamental frequency when talking to infants. As mentioned earlier, Ferguson (1977) suggested *Baby Talk* as a label for the systematic adjustments made by parents in Western cultures. The most striking feature of this variety is its high pitch, the so-called "nursery tone" and exaggerated pitch variations.

4.4 Visual communication

Human visual expressions are found for example in facial displays and body postures, but also in small movements of the muscles in face, hands and body, and finally by the colour of the face and neck. We will discuss each of these below.

4.4.1 Body postures

The body, primarily the way a person is walking, standing or sitting conveys at lot of information of the individual's background and present state of mind. Also a motionless body posture reveals emotional status, at least for emotions such as anger, surprise and happiness (Coulson 2004). As mentioned above, a large body signals dominance and a small body signals non-dominance. One way to express dominance is simply to stand up when talking to somebody sitting down. An erect position also signals physical and psychological balance, whereas the opposite – stooping shoulders and lowered head – gives the impression of a tired or "depressed" person. These behaviours are conspicuous in greeting behaviours where there are different ritualized patterns, where people decrease their size by nodding, bowing, or curtseying.

In a study from the British school context, Neill (1986) asked school children to rate teachers on the basis of body postures. The children were shown pictures drawn from a film of authentic classroom teaching. They were able to draw conclusions about the teachers without information about movements, tone of voice, or content of the actual lesson. They judged the teacher standing in an assured position, with one hand on her hip (this increases the shoulder size) as a friendly and helpful teacher, whereas the teacher who was holding on to the blackboard was judged as inefficient and unhelpful. These judgements seem to be based, at least partly, on the impression of shoulder size.

4.4.2 Hand movements – pointing

The hands are used for various functions during social interaction. They are used in sign language, as so-called emblems substituting words, for pointing, and to show rhythm when speaking. The hands are also much used in ritual greetings, in self-grooming or allo-grooming.

In this section we will first focus on pointing. Pointing has been studied from many aspects, for example with respect to cognitive functions, relation to target, and cultural differences in the configuration of the hand in the gesture. More recently, studies of animals understanding human pointing, have attracted a lot of interest (see Chapters 5 and 6).

In contrast to their early fascination of voices and faces, newborn children do not pay any attention to hand gestures in the beginning. If the mother points, the infant tends to look at the hand instead of the direction of the hand. Not until at the age of around one year, children comprehend the meaning of pointing gestures. This is taken as evidence of cognitive maturation, and there is all reason to assume an interconnection between pointing and the first words. Pointing is a manifestation of shared attention and referential communication, and it is used a lot in parent-child interaction, particularly to label objects in the environment. It involves *triadic* communication; between two individuals and an external referent. The child's own hand movements are first used to get attention, secondly to request something (imperative pointing), and thirdly, around the age of 12 months, pointing may be used also to direct other's attention to an external referent (declarative pointing). The fact that they point more when actively interacting with adults, than when the adult look away, shows it is a matter of sharing attention with others (Liszkowski et al. 2007).

In analyses of pointing, there is a differentiation between *proximal*, when the target is touched, or *distal*, when the target is remotely situated. Sometimes their hand configurations differ. In proximal pointing the index finger is used, and in distal pointing the whole hand is used. The perception of pointing is usually accurate, and few deviations are found, even if the pointing arm is partially blocked (Bangerter & Oppenheimer 2006).

In Western contexts, pointing with the index finger extended towards an object and the other fingers loosely or tightly curved is thought of as the most natural way to do it. Empirical studies of pointing across cultures, however, have shown that even if shared attention is used among all human cultures, the manner of how to accomplish this differs. For example, Kita and Essegby (2001) report that in Ghana it is only possible to point with the right hand, and that pointing with the left hand is unacceptable. This is connected to the whole concept of left, which in many cultures relate to something wrong. They studied how people gave route directions, and found a tendency to hide the left hand behind the back, to avoid it from being seen by the listener. Not all cultures use the hand or fingers in pointing. Instead gaze and lips are used for pointing, sometimes in combination with hand movements. The culturally different kinds of pointing behaviour have to be learned. When coming into contact with cultures with other pointing preferences it is not only a new language that has to be learned, but also how to point.

The most commonly reported function of pointing is to indicate some object or phenomenon in a particular place. However, it may also indicate the time dimension. In order to specify a certain time, speakers of the Australian language Guugu Yimithirr point to the part of the sky where the sun is visible at that time of the day (Haviland 2000). This is an interesting example of how also humans may use a geocentric system (cf. Chapter 2.9 on honey bees). In many cultures time is conceptualized as something that flows by, and speakers place themselves in relation to time, referring to their own body as the axis instead of using the angle of the sun. In English and many other languages, speakers point (often with the thumb) over their shoulder to indicate that something happened in the past, and they point forward with the hand if they refer to the future. The opposite pattern is found in Aymara, spoken in the Andean highlands. There, the speakers think of the future as being behind them and the past in front of them and consequently they gesture over the shoulder when referring to the future (Núnes & Sweetser 2006).

Pointing is often seen as a typical human invention. However, studies of other species have shown that referential pointing is comprehended, and sometimes also produced, in other species, for example nonhuman primates and dogs. There are some differences in the way the pointing is used. Human pointing is typically triadic (point to third entity), and distal (not touching), whereas other species may use dyadic (point to other or self), and proximal (touching the target) pointing (e.g. Tomasello & Camaioni 1997; for triadic pointing in dogs, see Chapter 6). Another distinction is that between imparative and declarative pointing. While both humans and animals seem to use imperative pointing (at least understand it), only humans use declarative pointing. Box 4.1 summarizes some of the analytic units that are used in descriptions of pointing by hand gestures. They are described according to function and form, and additional information about specific meanings and/or cultural constraints.

Box 4.1 Analytic units to describe functions and forms of pointing as hand or arm gesture

ASPECT		TYPE OF POINTING	ADDITIONAL INFORMATION
Function		imperative	aiming towards a desired object
		declarative, referential	shared attention, triadic communication
Form	1. relation to target	proximal	touching target
		distal	not touching target
	2. hand/arm configuration	only right hand/arm	use of left hand/arm is taboo in some cultures
		whole-hand point	less specific, often without gaze
		index finger	specific target, often with gaze
		middle finger	taboo in some cultures
		thumb	often for time reference

4.4.3 Other hand gestures

Hand gestures are often used in greeting ceremonies. There are many ways to use the hands: they may be given to somebody in a handshake; hands may be waved, raised, used in military salute. Waving is the gesture that is used first in young children. Common to all hand greetings is that an open hand indicates friendliness (or at least that there is no weapon hidden in the right hand). The military salute is assumed to have developed from the movement when the helmet visor in the suit of armour was opened and the knight showed that his feelings were friendly. These gestures are conventionalized and have to be taught to children. The importance of arranging the greeting behaviours was highlighted by the American sociologist Goffman (1959, 1967) in his classical investigations on interactions in face-to-face situations. Goffman claims that greeting behaviours function as an opening to the continuation of an interrupted relationship and to keep down the inter-personal aggression that could have been the result of the interruption. Farewells, on the other hand, close the interaction and prepare the participants for what they may expect next time they meet.

Box 4.2

GOFFMAN ON THE MEANINGS OF GREETINGS

The enthusiasm of greetings compensates for the weakening of the relationship caused by the absence just terminated, while the enthusiasm of farewells compensates the relationship for the harm that is about to be done to it by separation. (Goffman 1967: 41)

The hands can be used both for verbal and for nonverbal communication. An example of verbal gestural communication is the sign language used by the deaf community (see 3.6). Sign languages exhibit all the characteristics of other verbal languages, thay have to be learned, and they have lexicon and grammar. A difference is that words seem to have more of an iconic relationship to the referent than is the case for spoken languages.

Hand movements are also used in connection to speech production. Typically, listeners keep their hands still, while speakers gesture, and whenever there is a switch between speakers, there is also a change in who is moving the hands. Interestingly, speakers tend to gesture even when there is nobody watching, such as in a telephone conversation, or in complete darkness.

4.4.4 Head and face

The face is by all comparisons the most important site for communication. We recognize each other by the facial features and in passports it is usually enough with a photograph of the face for identification. Infants show a strong interest in the human face already from birth, in particular in the eye and mouth region. They readily smile at the sight of two dots painted on a cardboard, giving the impression of eyes (Ahrens 1954). Babies are found to mirror mouth opening and tongue protrusion already in the first day of life (Meltzoff 1981).

The angle of the head gives information of the emotional state. Darwin (1872/1965) proposed the bowed head to be a sign of shame or sadness, whereas the opposite, the erect head posture was linked to feelings of pride. Assuming that shame is connected to submission and pride to dominance the function of different head angles can be extended to display dominance or appeasement. In an experiment where subjects were asked to judge emotions from pictures of faces, Mignault and Chadhuri (2003) confirmed Darwin's claim that a bowed head indicates submission and raised head expresses dominance. Furthermore, their subjects correlated the bowed head to smiling, suggesting that head posture and smile work together to express appeasement. Other gestures involving the head are for example movements of nodding, wagging and shaking of the head. These expressions do not have the same meanings in all cultures, which can lead to serious misunderstandings. The nod generally means "yes" in Northern European countries, but the opposite, namely "no" for example in Greece.

The facial expression that makes the strongest impression on others is the smile. To smile is a friendly social act, which is also highly contagious. What happens in a smile is that the upper lip is raised and the mouth corners are retracted, which causes the teeth to be shown. This has intrigued many researchers, since bared teeth are usually expressions of threat, coming from an intention to bite. There are two different suggestions about the origin of the human smile. The biologist Jan van Hooff (1972) suggests that it is a visual expression, related to the baring of the teeth that is used by apes in greeting situations. The linguist John Ohala (1984), on the other hand, claims that the smile is primarily an acoustic signal which works according to the well-known 'frequency code'. When the corners of the mouth are drawn backwards, the resonance space in the mouth diminishes, and the sound that arises is a whining sound in a falsetto tone. This is exactly the kind of sound that would come from a small individual. The opposite pattern occurs in aggressive behaviour. Then the teeth are bared, but with the difference that the corners of the mouth are drawn forward, to increase the resonance space and the vocalisation becomes associated with threats from a larger individual. According to this explanation, the smile intends to decrease aggression by mimicking a smaller individual (this is, by the way, one of the features that humans share with some other species, e.g. the wolf).

Facial expressions are complex actions that are created by a combination of several different muscle groups. As mentioned above, it is possible to link certain muscle activities with specific expressions. However, it has not yet been possible to isolate which activity that gives a meaningful facial expression. The smile involves not only muscles around the mouth, but also the cheeks and the eyes, and therefore a more holistic approach is necessary. A smile that is produced with only mouth muscles can therefore be described as "not reaching the eyes" which means that it is not perceived as genuine. Genuine smiles (labelled the "Duchenne smile" by Ekman and Friesen, see below) involving the eye region are used selectively by infants towards family members, as soon as they are able to discriminate persons in the environment (Fox & Davidson 1988). The smiling behaviour is assumed to be an innate expression, but how and when to use it is subjected to some learning. Since smiling is also used for social interaction, it does not necessarily indicate a specific emotional state. People smile when they are happy, but also learn to give social smiles without feeling happy, as part of general social behaviour. Smiling is strongly contagious, which can be seen in early infancy. The synchronization, or imitative behaviour, that have been found in infants already from their first days in life is probably due to the triggering of the neurological mirror mechanism (Rizzolatti et al. 2002). This basic social feature may explain why adopted children often resemble the family they are adopted into (for example, they have the "father's smile"). These children have imitated a particular way of smiling from an early age and use it also in other contexts.

A comparison between facial muscle movements in humans and other primates reveals many similarities, but also some differences. One difference is in the upper part of the face, which seems to be specialized for eyebrow movements in humans (Vick et al. 2007, Waller et al. 2006). The eyebrows have a high communicative value. In situations of meeting and parting, there is a lot of intense eye contact and the raising of eyebrows plays a particularly important role as a reinforcement of greeting. In a study of cross-cultural comparisons, Eibl-Eibesfeldt (1972) used a method to film social interactions without people being aware of it, by using a mirror prism. He found the *eyebrow-flash* as a greeting signal occurring in all cultures that he investigated (for example in Bali, New Guinea and France), and it is therefore regarded as one of the universals in human communication. Sometimes eyebrow raising is even found as the sole marker of greeting, especially at a distance, sometimes it is accompanied by smiles, nods and verbal expressions.

The eyebrow flash is used in friendly contacts, but it is not regarded as appropriate behaviour between strangers and it is even suppressed and considered as indecent in Japan. In some cultures it has an extended use in flirting, asking for information and as a general sign of approval. "It is a "yes" to a social interaction" (Eibl-Eibesfeldt 1972: 299). The colouring of the eyebrows and the upper eyelid by women is part of the general attention to this facial region and makes the expression the more prominent.

The correlation between the appearance of facial expressions and the emotions they convey has been discussed. Basic emotions such as fear, anger, surprise and happiness seem to be expressed in much the same way across different cultures. Darwin (1872/1965) suggested that the expressions could be interpreted as intentional movements. For example anger could be traced back to an aggressive biting behaviour, with the eyes narrowed for protection from an attack from an antagonist. Another explanation is that opposite emotional expressions relate to social adaptation to the dimension helpless infant versus mature adult. When exposing informants to photographs of faces expressing the emotions anger and fear, Marsh et al. (2005) found that faces with baby-like features tended to be associated with fear, whereas mature faces were associated with anger. According to Konrad Lorenz (1943), there are "releasing stimuli" that prepare for care-giving behaviour, and hinder aggression. Baby-like features such as round faces and large round eyes are appeasing, whereas adult mature faces with small eyes, low brows and large jaws are threatening.

As mentioned above, the lips may be used as a pointing gesture. Studies from pointing behaviour in Africa, Australia, Oceania, and Southeast Asia have demonstrated that other body parts may be used, for example the upper or lower lips, the tongue or gaze. Kegl (2004) reports lip pointing in Nicaraguan Sign Language and concludes that use of lip pointing allows the hands to be used for other expressions. Lip pointing may differ slightly with respect to exactly how the lips are moved, sometimes both lips are used, sometimes only the upper, or lower lip.

4.4.5 Eyes and gaze

In comparison to nonhuman primates, the human face seems to be constructed in a way as to highlight the eyes. The conspicuous eye-brows give a frame to the eyes and the high cheekbones allow for eye muscle activity (Emery 2000). Furthermore, the human eye has a well-developed white area (sclera) surrounding the iris, which makes it different from eyes in other primates (and other mammals). The eye outline is extended horizontally and the movements of the eyeball are easy to detect (Kobayashi & Kohshima 2001). Taken together, these features contribute to make eyes and gaze important means for communication.

Research on the communicative expressions of the eye has mainly focussed on two aspects; the pupil size and the gaze pattern. One characteristic of the pupil is that it can vary in size under different conditions, both as a function of how much light there is in the environment, and as a function of inner excitement. When we see something interesting, the pupil is widened and when something unpleasant is shown, the pupil gets smaller. Poker players prefer to use dark glasses in order not to give away information about their hand by pupil size, and jewel dealers do the same to hide their excitement for a good prize. Hess (1972) conducted a study with manipulated pupils and showed that the size of the pupils functions as a clear signal. When shown two identical photos of a young woman where the pupils are enlarged on one photo but not on the other, young men found the woman with large pupils more attractive. There was also a positive pupil reaction from the part of the person who looked at the picture with the attractive woman. Other studies have shown similar results, for example when mothers are presented pictures of babies; they prefer the baby with large pupils before the baby with small pupils. The trick with manipulating eyes in pictures is also used in advertising, where models tend to have large pupils.

Eye gaze behaviour plays an important role in communication as a means to establish joint attention. Infants have been found to follow the gaze of an adult already at 9 months of age (Carpenter et al. 1998). In Western cultures mothers typically follow the infant's gaze and often comment what the infant is looking at. This is part of the growing triadic communication. Another gaze behaviour is the mutual gaze, which is used in face-to-face contacts, with or without speech. When the acoustic channel is used, one may expect it to be unnecessary to look at the speaker, but instead this seems to be of vital importance. Generally women tend to look more than men in verbal interactions. Listeners look more than the speaker, who may look away for a while and then look at the listener at the end of the utterance, which then is used as a turn-giving signal. In studies of group interaction in stable groups, it has been found that the leader is recognizable by the number of gazes from the others (Granström 1992). As we will see in Chapter 6, this is just like the alpha wolf in a wolf pack. Obviously, it is important to keep an eye on the leader and decode the emotions and attitudes.

In situations when the cognitive demands are high and planning arguments or find words take a lot of energy speakers tend to avoid eye contact instead. When the desired word is found the speaker may look at the listener again. It is assumed that planning an utterance is not possible to combine with interpreting the information that is coming from the facial expressions of the audience. This also implies that if a suspect looks away when accounting for his alibi this is not necessarily to be interpreted as a bad conscience, it may as well be utterance planning or word-finding problems that demand the whole attention (however, it is also the case that people do look down when they are ashamed or embarrassed; Darwin 1872/1965).

There are occasions when visual contact is not desired. Goffman (1963) talks about "civil inattention". In situations when strangers have to share a narrow space, for example in a lift or at a bus, it seems necessary to avoid eye contact. Instead the focus is directed towards the ceiling or the floor. Sometimes there is a quick glance showing that one is aware of the other person's presence, but after that the gaze is averted. If it is impossible to avoid having eye contact there is often a smile or a verbal utterance. The direct look at a stranger without saying anything may be interpreted as a threat.

However, there are also contexts when a steady eye-to-eye contact is not perceived as agonistic, but the direct opposite of aggression – this is the loving interaction between caretaker and children, or between lovers (Robson 1967, Berko Gleason 1977, Papousek & Papousek 1977, Trevarthen 1977). During breast-feeding the infant searches to the mother's eye-contact (Bowlby 1973), and mothers tend to hold the baby in a position *en face* (Klaus et al. 1970).

In contrast to pupil reactions, eye gaze is assumed to be culturally determined and the patterns described above, found in North European and North American studies, are not the same in all cultures. Some cultures demand an intense gaze in situations where it would be impolite in other cultures. In some cultures children are taught not to look directly at their teachers and other adults. In other cultures a steady gaze is a sign of confidence. Differences like these are often the source of misunderstandings.

4.4.6 Complexion

The facial skin colour conveys information about blood circulation, hormonal status and health. Fink et al. (2006) investigated how people rate photographs of female faces according to health, youth and attractiveness. 430 participants were asked to rate faces using a 10-point scale. The results demonstrated that the colour of the face has a strong impact of the perception of the female as healthy, young and attractive. Both males and females participated in the study, and came to similar results. The only difference was that males found it significantly easier to perform the rating task. This suggests that the association between colour and mate choice found in birds, butterflies, fish and nonhuman primates, also applies to humans.

The colour of the skin can also change according to inner emotions, particularly in the face and neck regions. Fear may result in a pale face, and anger and embarrassment in a red colour in the face and around the neck. Darwin claimed that blushing was "the most human of all expressions" (p. 153). He also noted that women blush more than men, and children more than adults, but not until the child had reached an age of self-consciousness. The colouring is described as signs of inner emotions in the literature, for example when Jane Austen describes the displays of emotion in Mr. Darcy and Elizabeth Bennet after he has proposed to her in (in Pride and Prejudice, 1813)."Elizabeth's astonishment was beyond expression. She stared, coloured, doubted, and was silent."

4.5 Tactile communication

The human body has a large area (1,5 + 2 m²) of pressure detectors and hair cells by which tactile signals may be received. However, human tactile communication seems to be restricted to very young children and between lovers. In young children, touching is used in tickling games and wrestling, but as children grow older the amount of touching decreases, until around puberty, when there is a new increase in touching. Tactile communication plays an important role in behaviours having to do with greetings and farewells, congratulations, and different types of ceremonies. Touch can easily be observed at places where people meet and part, for example in airports and railway stations. In a study on behaviour at American airports, Greenbaum and Rosenfeld (1980) detected 103 tactile behaviours, which they classified into the following types: kiss on mouth/cheek; touch on head/arm/back, embrace, and holding/shaking hands. They also found a gender difference, with females typically kissing and embracing and males typically shaking hands. There are different ways of handshaking. Brosnahan (1979) describes handshaking according to position of hand, pressure, fingers, and use of the left hand. By use of these parameters, she was able to capture certain gender differences. For example, male-male encounters were characterized by harder hand pressure, whereas female-female encounters were characterized by more of both hands (with the left hand placed on the upper arm or wrist of the other person).

There are many cultural differences in tactile communication. Studies of European tactile communication have shown that there is a tendency for countries around the Mediterranean to use touch more than others. For example, Remland et al. (1995) studied interpersonal behaviour at public places (railway/bus station) in the Netherlands, France, England, Ireland, Scotland, Greece and Italy. The material consisted of 381 situations where people met (so called dyads). The results showed a variation from social behaviour in Greece, where 32% of the dyads exhibited touching, to the Dutch dyads, with only 4% touching (Remland et al. 1995). This confirms the view of Greece and Italian as contact cultures and England, France and the Netherlands as noncontact cultures. France did not follow the expected pattern in this study, maybe due to the fact that the data was collected in central France (Lyon and Chartres). The authors suggest that French people living closer to the Mediterranean might have behaved more like the Italian and Greek informants.

Many social settings involve interactions between strangers. The use and effect of tactile communication on strangers was investigated in a study in a shopping centre, where the interviewer stopped people and asked them to fill in a form (Paulsell & Goldman 1984). When handing over the forms, the interviewer dropped them on the ground, seemingly by accident. If the interviewer touched the person that was stopped, there was a greater chance for that person to help in picking up the forms. The highest chances were if the upper arm was touched, next came touching of the lower arm. Touching the shoulder or the hand did not give any effect.

Interpersonal distance, i.e. how people place themselves in relation to others is another research area related to tactile communication. The American anthropologist Edward Hall (1966, 1975) coined the term "proxemics" for the use of space. He identified four interpersonal distances: the intimate distance between lovers, the personal distance between friends, the social distance at business meetings, and public distance on public occasions. These distances differ between cultures, so that the intimate distance in one culture may be the same as the personal distance in another culture, thereby causing conflict situations. Cultural differences can therefore be observed by the way people take their places on a bus, or on a public bench, and by how tables and chairs are organized in cafes and restaurants.

4.6   Chemical communication

Chemical signals are received by the olfactory system, which consists of cells in the nose and connected regions in the brain. The olfactory expressions have been investigated to a lesser degree than other communicative channels. However, the groundbreaking studies by Linda Buck and Rickard Axel (who were rewarded the 2004 Nobel Prize in Physiology or Medicine) gave new insights about the olfactory system. They showed that the smell stimuli are processed in similar way to visual stimuli, with the brain creating an odour representation from a combination of different components (Buck & Axel 1991, Buck 2000, Zou & Buck 2006).

Humans are generally not very good at distinguishing between different odours, and even quite familiar smells, like coffee, chocolate and orange are difficult to differentiate and name (Jönsson et al. 2005). In fact, of the 1000 human olfactory genes, about two-third are assumed to have been made ineffective by different mutations (Stockhorst & Pietrowsky 2004). The olfactory signal is highly complex and each specific smell can be analysed into different volatives. For example, the odour of coffee consists of more than 800 volatives (Hudson 1999). By and large, the communicative value of human olfactory signals is unclear, and hard to investigate experimentally. However, the fact that human body odours are being eliminated in many cultures, and replaced by perfumes (sometimes based on animal secretion, for example musk from the musk deer) reveals that olfactory signals are important for us.

Studies of the area of human sexual interaction, have found the olfactory signalling by way of *pheromones* to be of importance. Pheromones are not consciously detected but can be observed only by their arousal that they activate, which is generally associated to the area of sexual and reproductive behaviours (cf. 2.7). Research on human chemical communication has often focused on the upper body, for example by presenting a pad with the odours of the mother's breast to a newborn baby and showing T-shirts to adults for identification. In both cases, there is evidence showing individual recognition (Wyatt 2003). Several studies show that human pheromones are used to coordinate behaviour, since for example females living together in a dormitory have been found to synchronize their menstrual cycles (McClintock 1971).

Another result is that women tend to prefer the smell of men that are genetically dissimilar, which may serve the important function of avoiding inbreeding. In a study where subjects were asked to assess attractiveness of the smells of t-shirt worn by similar and dissimilar individuals, it was found that women rated the dissimilar men higher (Wedekind et al. 1995).

Box 4.3 Overview of some studies on human nonverbal communication

CHANNEL	MEDIUM	EXAMPLE	INFORMATION (PRIMARILY) ABOUT
Acoustic	Voice (extralinguistics)	Laughter, weeping	Emotions (Owren & Bachoroski 2003, Nelson 2005)
	Voice (paralinguistics)	Speech rate, pitch, pauses, loudness	Emotions (Scherer 2003), age (Schötz 2006), identity (Zetterholm 2003), listener (Ferguson 1977)
Visual	Body posture	Slooping shoulders	Emotions (Coulson 2004), personality (Neill 1986)
	Hands	Gestures	External referents (Kita 2003, McNeill 1992)
	Head, face	Smiles, frowns, eyebrow flash	Emotions (Ekman & Friesen 1978, Hendriks & Vingerhoets 2006), social relations (Eibl-Eibesfeldt 1972)
	Eyes, gaze	Pupil dilation	Interest (Hess 1972)
		Gaze	Attitudes, external referents (Argyle & Cook 1976)
	Complexion	Colour	Physiological condition (Fink et al. 2006)
Tactile	Body	Hugs, kisses, beats	Attitudes, social relations (Paulsell & Goodman 1984)
	Hands	Handshaking	Social relations (Schiffrin 1974)
		Touching	Attitudes (Paulsell & Goodman 1984)
	Proxemics	Distance	Social relations (Hall 1966)
Chemical	Glands	Sweating	Physiological condition, emotion (Wedekind et al. 1995)
		Lactating mothers	Identity (Wyatt 2003)

4.7 A model for analyzing gestures – The Ekman and Friesen model

One of the most influential models for the analyses and categorization of nonverbal communication is Ekman and Friesen's now classical article from 1969, "The Repertoire of Nonverbal Behaviour: Categories, Origins, Usage and Coding". The suggested categories are based on Efron (1941) and include emblems, illustrators, regulators, affect displays and adaptors. These categories can be seen as points along a continuum from intentional, learned, culture-specific expressions to unintentional, innate, and universal expressions according to the following:

Box 4.4 Continuum of nonverbal categories, modified after Ekman and Friesen (1969)

Emblem	Illustrator	Regulator	Affect display	Adaptor
←————————————————————————————————→				
intentional				unintentional
learned				innate
culture-specific				universal

Emblems are intentional nonverbal gestures that have verbal translations. That there is an intention involved is clear from the fact that emblems are usually combined with eye contact. One example is the "Thumb-up" sign, assumed to derive from the Roman games with lions, where the emperor would show with the thumb whether the prisoner was to be liberated or not. Another example is the "V-sign", commonly associated with "victory" when the palm is directed outwards. Often the verbal translation is taboo, and therefore a gesture is used (which also, of course may be taboo but easier to hide!). Which gestures are taboo differ between cultures. One example is the V-sign, which has different meaning in different cultures (Morris et al. 1979). In other words, meanings of emblems are shared across individuals in a group much in the same way that meaning of words. The fact that people are aware of how they use emblems makes it easier to investigate them, in comparison to studying other forms of nonverbal behaviour, that are unplanned and with a low degree of awareness. This is also the reason why nonverbal communication often is exemplified by means of emblems – they are salient and what people remember.

Some emblems are arbitrary, i.e. it is difficult to find a relation between the gesture and its meaning, e.g. the "finger-ring sign" (meaning 'Okay' in some cultures), other may be more or less iconic, i.e. with a clear association to what they stand for, for example, when shaking the fist to show aggression, which can be interpreted as an intention to hit. Ekman and Friesen suggest that pan-cultural emblems (i.e. the ones denoting the same meaning in different cultures) are those, which refer to primary body functions such as eating, drinking and sleeping. Other emblems refer to more complex human activities, where cultural institutions are involved, for example pretending to hold a telephone receiver.

Illustrators are gestures that supplement the verbal communication and are used simultaneously with speech. The types of illustrators identified by Ekman and Friesen are batons, deictic illustrators, ideographs, pictographs, kinetographs and spatial movements. These have slightly different conditions; the batons and ideographs accompany verbal utterances, whereas deictic illustrators may be used also independently of speech. Batons, or beats, are hand gestures with no specific meanings; they are used to segment the verbal speech stream, stress what is important in the message, and also to help the speaker to keep the turn. As long as the speaker moves the hands, others will not interrupt. Typically, when people have finished talking they tend to put their hands to rest in the lap. Deictic illustrators involve pointing and are easily discernible from other gestures. They are referential in that they direct the attention to an external referent, e.g. person, object, direction, or location. Ideographs, pictographs, kinetographs and spatial movements are more or less iconic in the sense that they resemble the concept they are referring to. For example, illustrating the size and shape of an object (pictograph), depict a bodily movement (kinetograph), or the course of a thought (ideograph).

Regulators are defined as "acts which maintain and regulate the back-and-forth nature of speaking and listening between two or more interactants" (Ekman & Friesen 1969: 82). Turn taking is signalled in many ways, for example by head nods, eye contacts, eyebrow movements, postural shifts and vocal feedback signals such as mm-hmm. Typically speakers gaze less than listeners, but at the transition point, when the roles of speaker and listener are to be switched, the speakers tend to look directly at the eyes of the listener. This expression is often combined with a lowering of the hands and laying them to rest. Acoustic features such as pitch changes and a drawl on the last syllable also prepares for the turn taking switch.

Many *affect displays*, for example smiling, crying and blushing, are assumed to be innate. This can be evidenced by studies of children who are born blind and deaf and still smile the same way as other children. What is culturally learned is often how to suppress the behaviour. Pictures of sportsmen succeeding in championships typically show men and women with arms up in the air, laughing and smiling. However, it is obvious that different cultures allow more or less affect displays. Children early learn to control certain behaviours, and learn which affect displays are appropriate in which situations.

Adaptors have been described as fragments of behaviour that was used in childhood, but is reduced under cultural pressure. Adaptors are not used intentionally as signals, but may be interpreted as signals, revealing emotions to the person displaying them. Adaptors seldom generate any feedback behaviour from the conversational partners, except for the synchronization movements that may occur between friends in a group.

In those cases, an adaptor may trigger a comparable adaptor in others – so if A scratches her head, B does the same. Ekman and Friesen identify three types of adaptors: self-adaptors, alter-adaptors and object-adaptors. Self-adaptors are grooming-like behaviours that are directed towards one's own body, whereas alter-adaptors involve touching other persons, and object-adaptors involve touching and manipulating objects.

Adaptors can be seen as what Tinbergen (1960) called displacement activities, when discussing similar behaviour in animals. They function as outlet of nervousness or aggression and typically involve cleaning or grooming movements, for example a bird wipes its beak, a cat licks itself, a man strokes his beard, and a woman adjusts her hair. There is an advantage to have cleaning behaviour latent, as long as it does not go too far, which is the case for example in nail-biting. Adaptors are innate and thus generally similar across cultures, except for some behaviours dealing with cultural artefacts, such as biting on a ballpoint pen, or culturally unwanted behaviours such as cleaning the teeth or nose.

The Ekman and Friesen model has been proven a successful model to categorize and describe behaviour in normal and pathological individuals. It has intuitive appeal and the categories are simple enough to be applied in therapy and research without specific training. This shows that people in general are able to generalize communicative expressions by recognizing these holistic patterns. However, there are many details of the nonverbal communicative behaviour in humans that are not captured in the model. Therefore we will also give another perspective on gestures: the speech-related gestures.

4.8   Where verbal and non-verbal communication meet: Speech-related gestures

During the last decennia, many have questioned the earlier division of verbal and nonverbal communication into two separate systems. Instead, it has been suggested that some gestures are closer than others to speech, and therefore another analytic framework has been suggested in work by for example Kendon (1980, 2004), McNeill (1985, 1992, 2005) and Calbris (1990, 2011). They all argue that the components gestures, prosody and words work together to realize meanings.

Anyone observing people engaged in conversation soon becomes aware of the speaker moving the hands in rhythmic movements. The movements vary in intensity across cultures and between individuals in the same culture, but the common denominator is that they seem to follow the speech rhythm. The hand gestures occurring with vocalizations have their earliest expression in infants' babbling. Simultaneously with the emergence of repetitive babble (*dadada, mamama*) the child is able to move the hands or arms in a rhythmic manner (for example, by hammering the spoon on the table). The speaker does not use the hands at random, but move them in a very

Box 4.5

> **MCNEILL ON GESTURES AND SPEECH**
>
> There is no separate "gesture language" alongside of spoken language. Indeed, the important thing about gestures is that they are not fixed. They are free and reveal the idiosyncratic imagery of thought. Yet, at the same time, such gestures and the images behind them coexist with speech. They are tightly intertwined with spoken language in time, meaning and function; so closely linked are they that we should regard the gesture and the spoken utterance as different sides of a single underlying mental process. (McNeill 1992:1)

systematic way. Because of their close relationship with verbal language, gestures co-occurring with speech have been seen as fundamentally different from other kinds of nonverbal communication and different models have been developed for their analysis and to account for the exact form of the gesture-speech link. One proposal is that the gestures emerge from the pre-linguistic concept and therefore they are not influenced by the linguistic structure (e.g. Krauss et al. 2000). Another view is that gestures are parts of the verbal utterance and carry the psychological predicate (McNeill 1992).

4.8.1 The McNeill model

In his book "Hand and mind. What gestures reveal about thought" (1992) McNeill proposes a model for analysing gestures. The model deals with manual gestures, performed by arm, hand and fingers, and in conjunction with speech. This implies that

Box 4.6 Categories suggested by McNeill (1992)

> **Iconics** are gestures that depict an action or an object, for example by showing opening by a rotating movement of two hands.
>
> **Metaphorics** resemble iconics, but represent an abstract idea rather than an object. For example, grasping with the hand in the air to illustrate "we must deal with the situation".
>
> **Beats** are hand movements in the same rhythm as speech. They are not only used to mark the tempo, but also to introduce new themes in the utterance – typical in political debates.
>
> **Cohesives** serve the opposite function to beats. Instead of marking something new, they mark that the parts of the utterance are tied together. Political speeches usually show a large amount of cohesive gestures.
>
> **Deictics** indicate objects and referents, either present in the speech situation or abstract entities that have been introduced in the conversation. For example, when talking about several different persons, speakers may refer to them by placing them on an arena and then point at their respective place. (Gullberg 1998, Haviland 2003)

other nonverbal expressions, e.g. smiling, body postures are not taken into account. Nor are emblems considered, since they are typically used in the absence of speech. Five categories of gestures co-occurring with speech are proposed: iconics, metaphorics, beats, cohesives and deictics.

If gestures and speech are integrated, their timing should be a central feature. The gesture unit consists of several phases. First there is a preparation, then the stroke, and then a recovery. The synchronization of gesture and speech takes place after the preparation, so that the stroke coincides with the spoken word. This means that the preparation of the gesture actually takes place before the production of the word. In McNeill's words:

Box 4.7

> **MCNEILL ON CO-EXPRESSION**
>
> Co-expressive symbols, spoken and gestured are presented by the speaker at the same time – a single underlying idea in speech and gesture simultaneously. The synchrony is crucial, because it implies that, at the moment of speaking, the mind is doing the same thing in two ways, not two separate things, and this double essence is a reason for positing a dialectic of imagery and language.　　　　　　　　　　　　　　　(McNeill 2005:23)

The timing between gesture and speech is discussed in detail in Calbris (2011; with examples that can be studied in the video-clips online that accompany her book). The hypothesis is that the gestural expression is processed faster than the spoken expression. The gesturing helps the speaker to put words to the ideas. This also helps the interlocutor to anticipate what the speaker is going to say, and it actually involves the interlocutor in the utterance process. Furthermore, since gestures are often more concrete, they may serve as pedagogical tools, unintended by the speaker, but helpful to the interlocutor ".. the concrete gestural representation (…) may allow a less informed listener to guess the meaning of words unknown to him". (Calbris 2011: 254).

Summarizing, McNeill's model provides a conceptual framework for how to integrate the analyses of gesture and speech, and the categories suggested have had a considerable influence over subsequent research on gestures. However it does not explain why not all parts of an utterance are reflected by gestures, why in most speech situations there is more speech than gestures. McNeill (1992) suggests a one-gesture/one clause relation, but in an investigation of 21 subjects retelling cartoon stories, Beattie and Shovelton (2002) found only 86 iconic gestures in 1194 clauses, i.e. the large majority of clauses were not accompanied by such gestures. It is possible that the type of narrative decides how many gestures are to be used. A closer analysis of the Beattie and Shovelton data showed that it was more common to use gesture when an utterance contained a part of the story from a participant viewpoint than from an observer viewpoint. Furthermore, the events that were matched with gestures were unique and unfamiliar events, for example, when a dog put two baskets together in order to make a ball. This finding has implications for the interpretation of gesture function. It goes against the layman hypothesis that gestures are used when there is a word-finding problem, i.e. when the speaker cannot find the proper word. Instead, some gestures are used when the words are there but the content is in some way unexpected.

Most research has focused on the production of gestures, and less is known about exactly how the information given by gestures is interpreted by the addressees. Since information that is only expressed gesturally may turn up when informants are asked questions about a story, those gestures must have been noticed and processed (Beattie & Shovelton 1999a, b). But are all gestures noticed? Gullberg and Holmqvist (2006) investigated visual perception of gestures by using an eye-tracking technique, which indicates fixation points of the addressee. The informant listened to a retelling of a short narrative, equipped with an eye-tracker on the head. The analyses showed that most visual attention was given to the speaker's face, and that gestures were usually not fixated. However, in two contexts the listener would fixate the gestures: (1) if the speaker looked at the gesturing hands, and (2) when the hands came to a halt and stopped moving. The interpretation is that peripheral vision might be enough to see the gestures, and that the social norms in human interaction put restrictions on what a listener may be looking at.

In simultaneous translation, the gestural component of the talk is often lost (Weale 1997). Even though the interpreter is able to see the speaker and perceive the gestures, so much of the personality is expressed through way of talking, that it may be difficult for the interpreter to impersonate someone from another cultural background. Some gestures cannot be used, because they have different meanings in different cultures (e.g. the finger-ring gesture), others have no equivalent in the target language. There is also a time lag between the speaker and the interpreter, since the interpretation is not done word by word, but in a more holistic manner. Weale (1997) calls it "the art of the impossible", since the interpreter is at the same time playing another's role, his own role and attending to reactions from the audience.

4.8.2 Using knowledge of nonverbal expressions in verbal descriptions

As mentioned in Chapter 3, the advantage of verbal language is that it makes it possible to communicate events that are outside the immediate context of "here and now". To express inner emotions, on the other hand, we use the body: facial expression, gestures, touching, or the voice: prosody (intonation), cries, laughter etc. Even when describing emotions verbally, we tend to resort to nonverbal communication. Posture, movements, face and gestures are often used in metaphoric expressions. When we hear expressions such as *having no guts, to cold-shoulder someone, to have cold feet, to have a tongue-in-cheek approach*, we do not interpret them literally but as signs of personal characteristics, attitudes or emotions. For example, to lack "guts" or "spine" does not mean to actually lack them as body parts, but to lack what they signify, namely a strong character and will, expressed by an upright position.

A basic rule of metaphors is that positive concepts are associated with "up" and negative with "down" (Lakoff & Johnson 1980, Kövecses 2002). There is a link between bodily expression and metaphors, as "drooping posture typically goes along with sadness and depression, erect posture with a positive emotional state" (Lakoff & Johnson 1980: 15). Therefore, it should not come as a surprise if the same posture or part of the body is referred to when expressing the same emotions across different languages.

One striking fact about metaphors expressing emotions and attitudes is that they contain part-whole relationships. This means that we do not need detailed information of the body, but one significant part of the body may stand for the whole expression. For example, the nose is often used to illustrate arrogant behaviour; one is conceited both in English, French, Swedish and Russian by "turning one's nose up" (Italian: *con il naso all insù*; French: *marcher le nez levé*; Swedish: *sätta näsan i vädret*, Russian: *zadiraty nos*). The head may also be used as in French *garder la tête haute* (keep the head high), the same meaning as Hindi *shir uchcha rakna*. Sayers (1995: 372–373) gives examples on how emotional states are expressed in Wik-Mungkan, an Australian aboriginal language. Most descriptive expressions are compounds, made up of body parts and verbs. For instance, the word for nose is used for conceit (*kaa'thayan*, literally 'nose-up'); the eye is used to express surprise (*mee'ikan*) and the ear to express alertness (*kon-thayan*, literally 'ear-firm'). Sometimes there is a slight difference between languages in choice of body part to describe the emotion. To be depressed or disappointed is expressed in English by being 'down in the mouth', in Swedish by 'hanging the lips' (*hänga läpp*) and in Russian by 'letting the nose hang' (*veshaty nos*). To be sulking is expressed by 'blowing up the lip' in Russian (*naduty gubi*) and by 'showing a thick under lip' in French (*faire la lippe*). All the time it is the same facial expression that is described, but different parts of the face are chosen, the mouth, the lips, or the nose.

The tendency found among all social animals, to increase size in order to be threatening and to make them smaller to show submission, is also found behind many verbal expressions for emotion concepts. For example, in the Austronesian language Gidabal a conceited person is called *nyuladahn*, meaning 'very much him', in Tagalog a proud person is 'growing inside' *lumaki ang loob* and gets 'a big head' *lamaki ang ulo* (Palmer & Brown 1998). A big person has to look downward in order to look into other people's faces. In English this is metaphorically expressed as *Look down on somebody; Look down one's nose at somebody*. Contempt is expressed as *Turn one's nose up at someone*. If a person is *brought to his knees* he may *look crestfallen* and *crawl in the dust*.

Up until now we have only dealt with expressions of emotions that use metaphors to convey emotions and moods. There are also specific lexical items – adjectives, nouns and verbs – designed for emotional expressions. Certain adjectives, such as *glad, sorry, angry, afraid* are referred to as emotional adjectives. Thus, although the nonverbal channel can be considered the prime channel of emotive expressions, there are also verbal means. Interestingly enough, the etymologies of the emotional words often seem to refer to bodily behaviour, so that the words expressing inner states of mind are derived from exterior signs of these states. In fact, the body can be traced in the etymology of many adjectives, for example *glad* and *anxious*, and nouns such as *anger, wrath*, and *shame*. *Glad* belongs to a group of words beginning in gl- which all have the meaning of brightness and smoothness; glitter, glimmer, glossy etc. These words can all be associated with a happy and calm face. *Wrathful* is the opposite of *glad*. *Wrath* is related to the verb *writhe*, which means 'to squirm or twist the body about' (Random House Dictionary) and this is exactly what an wrathful face feels and looks like. *Anger* is related to Latin *angor* meaning 'contracted throat', which is another feeling of anger. *Shame* is derived from the Old Indo-European word *(s)kam*, 'cover', with the meaning that something is covered. This is typically expressed nonverbally with hands covering the face. Similar links between bodily expressions and adjective etymology can be found in other languages, for example a connection between trembling movements and fear, found in German *Schreck*, Hungarian *reszket*, Russian *bojat'sa*, and Swedish *rädsla* (Håkansson 1989, Györi 1998).

In literary works, where we cannot use tone, face or gestures to account for the characters' feelings and experiences, precise descriptions of nonverbal expression of emotions are particularly important. There are many different ways to describe the nonverbal dimension in narrative literature (see Poyatos 1997). For example, punctuation may be used to mark tone of voice indicating whether something is said as a question or exclamation (*Now? Now!*), focussed words may be marked by capital letters or by italics, repetition can indicate hesitation, or emphasis. The visual expressions are described by verbal descriptions, e.g. *her face flushed, she shook her head impatiently,* "*From shame and despair he hid his face in his hands*" (Tolstoy: Anna Karenina). By these descriptions, the inner emotions are described by depicting the nonverbal expression. When it comes to characterising vocal communication, there is a difference between using sound-reproducing verbs (such as *mutter, growl, snap, laugh*), descriptions where the verb is modified by adjectives or nouns (e.g. *whispered softly, spoke hoarsely, said with a grin/a shiver*) on the one hand, and onomatopoetic transcriptions on the other hand (*Whee-aa-oo!, Whooee! Ooooh-ooh, Ahem? Uh?*). Both types are symbolising the sound, and based on onomatopoetica. These transcriptions are of the same type as is used in descriptions of animal vocalizations (see Chapters 5–7) and they have the same drawback, namely that they give an impression of being faithful renderings of the sound, when they in fact are biased and based on the particular language of the person observing and describing the sound. Consequently, just like words, onomatopoetic descriptions have to be translated between languages. When Winnie-the-Pooh in his "bee-song" sings *Buzz-Buzz-Buzz* in English, he sings *Surr-Surr-Surr* in Swedish (Milne 1965/Swedish translation 1979). When he falls down, it goes *Crack!* in English, but *BRAK* in Swedish, when Cristopher Robin walks up the stairs, Winnie-the-Pooh goes *bump-bump-bump* the stairs in English and *duns-duns-duns* in Swedish. In an analysis of translations of nonverbal behaviour in Alice in Wonderland, Nord (1997) concludes that paralinguistic features are particularly difficult to translate, and it is sometimes better to omit expressive exclamations altogether, than to try to reproduce them.

4.9 When production is not automatized

From a layman perspective, it is often thought that gestures can compensate for production problems and help finding the right word. In this section, we will examine this assumption in three contexts, (1) when there is a problem with the oral production, resulting in stuttering; (2) in children with language impairments; and (3) in adults learning a second language.

Stuttering has a visual parallel in gesturing. In a study on the relationship between gesture and speech in patients with stutter problems, it was found that the gestures were interrupted during a stutter "the gesturing hand froze in midair during the stuttering bout" (Mayberry & Jacques 2000: 207).

Box 4.8 Stuttering gestures

> **MAYBERRY & JACQUES ON STUTTERING GESTURES**
>
> The speaker's mind coordinates the complex patterns of the gesture system (finger, hand, arm, shoulders and joint neuromuscular patterns) and integrates them with the complex patterns of the speech system (orofacial, laryngeal, and respiratory patterns) while weaving all this together with the complex mental structures of thought and language. The integration and timing the mind produces are so highly coordinated that all these vocal and gestural actions appear as one seamless image in the same moment of expressive time.
> (Mayberry & Jacques 2000: 212)

This example shows that manual gestures are not available as compensation for problems in oral production; rather that problems in one area, oral production, give problems in the other, the gesture.

Children develop gestures alongside speech. The earliest words and one-word utterances are often accompanied by pointing gestures. An interesting aspect of children's gesture production is that, in contrast to adults, where speech and gesture express the same (or related) information, children's gestures and speech may give different information. For example, the child may say "open" and point at a drawer, or say "bear" and scratch with the palm in the air, indicating the claws (Goldin-Meadow & Butcher 2003). The children are producing two-element utterances using two modalities. Another difference from adults is that young children have a larger gesture space and typically act out the gesture with their whole body or use oversized hand gestures (McNeill 1992). Children with language impairment may differ from normally developing children not only in their verbal expressions but also in their way of using gestures. Månsson (2003) found that children with language impairment used iconic gestures that were less complex, sometimes with only the perceptual but not the functional cues. For example, one child represented the concept of "binoculars" with one hand alternating from eye to eye, instead of having one hand put to each eye to perform the function of binoculars. The interpretation is that the problems in this group of children are present both in the verbal domain and in the nonverbal conceptualization, i.e. language problems are not compensated by gestures.

The same picture emerges in studies of second language learners. Gullberg (1998, 2006) asked learners to retell a cartoon story in both their first and their second language, and found that most informants (nine out of ten) used more gestures in the second language. However, when it came to the iconic gestures, these were used more in the L1 than in the L2 (Gullberg 1998, see also Nicoladis 2002). This shows that these gestures can only be accessed at the same level as words, i.e. when the speaker knows the word.

4.10 Summary

Nonverbal communication is an inseparable part of human communication. It is used to express affective and social functions outside of language, but also to optimize verbal communication. Emotions, attitudes and interpersonal relations are typically communicated by using acoustic, visual, tactile, and chemical expressions outside of speech. We often find strong similarities between human nonverbal communication and communication in other social animals. For example sympathy can be signalled through synchronization of movements and postures. However, gestures are important also in verbal communication, Deictic gestures with hand or gaze are crucial for human communication, creating a space for joint attention, and allowing for referential communication.

4.11 Suggested readings

Calbris, G. (2011). Elements of Meaning in Gesture. Amsterdam: Benjamins.
Calbris presents a thorough account of gestures as a symbolic system. The discussion is illustrated with a large variety of examples, taken from everyday conversations. Videoclips of some of the examples can be found online, which help giving the reader a full understanding of how gestures and speech together contribute to the meaning.

Goffman, E. (1967). Interaction Ritual. Essays on Face-to Face Behaviour. New York: Anchor Books.
This is a collection of essays which has become classic reading on human interaction. Goffman describes how glances, gestures and postures are used together with language in everyday meetings, on the street, on a bus, at a work place. A central concept is "Face-work" – how we use different strategies in order to "save our face".

Kendon, A. (2004). Gesture. Visible action as utterance. Cambridge: Cambridge University Press.
This book, written by one of the leading experts in the field, gives an overview of gestures, starting in the interest of gestures in the Classical Antiquity and covering the field till how gestures are classified in the 21st century. Kendon uses own video illustrations together with transcripts of speech to illustrate gestures used in everyday interaction. The illustrations, together with discussion of definitions and analyses make this book a valuable textbook for all students of gestures.

Chapter 5

Our closest relatives – nonhuman primates

5.1 Introduction

Communication among our closest relatives is both similar and different from human communication. It is similar because we share a repertoire of body postures and movements, hand gestures and face mimics, but it is different when it comes to vocalizations and communicative behaviour. To give a simplified picture: in a group of humans there is a lot of vocalizations, and people place themselves face-to-face, whereas the individuals in a group of nonhuman primates socialize by turning their backs to each other and involve in grooming. This chapter will describe communications in nonhuman primates from what has been observed in a number of studies of the last century, and also present some experiments where human language has been taught to nonhuman primates. But – who are the primates?

5.1.1 The primate family

In his classification system, Linneaus (1758) suggested one group of mammals to be called *primates*, since he regarded it being superior to other animals. The primate order is large, and divided into several families and species. Humans belong to the group classified as Old World monkeys (i.e. primates in Africa and Asia), more precisely to the *Hominoidea* family. This family includes for example gibbons (*Hylobates lar*), orangutangs (*Pongo pygmaeus*), gorillas (*Gorilla gorilla*), chimpanzees, (*Pan troglodytes*: with three subspecies), bonobos (*Pan paniscus*), and humans (*Homo sapiens*).

During history there were several splits, so that first the gibbons formed other lineages, whereas orangutangs, gorillas, chimpanzees, bonobos and humans formed the group of Great Apes. The chimpanzee and the bonobo are our closest living relatives. In fact, the chimpanzee is even closer to humans than to gorillas. The relationship is based on genetic evidence, which suggests that we share about 99% of our genes (King & Wilson 1975). Many morphological features are shared between humans and other members of the family, for example: no tail, five fingers, two nipples, and a binocular vision. Among characteristics that are not shared, are brain size and bi-pedal walking. Compared to nonhuman primates, humans have a larger brain size relative to body size and only humans use bipedalism as the normal way of transportation. When it comes to behaviour, one, maybe the most important, difference between humans and nonhuman primates is the communicative behaviour.

Studies on communication in nonhuman primates have often focussed on the issue of referential/informative function and intentionality – that is, they have used the sender-message-receiver paradigm. However, there are also studies indicating resemblances between human and non-human primates when it comes to synchronization of vocalizations and gestures. This has clear social functions to make interaction and cooperation easy.

In this chapter, we will mostly talk about the Great Apes, but we will also give some examples from other Old World monkeys (baboons, vervet monkeys, Diana monkeys, Campbell's monkeys and macaques) and some New World monkeys (i.e. from South and Central America; for example capuchins, squirrel monkeys, tamarins and marmosets). As we will see, when it comes to communicative behaviour some of our distant relatives are far more like us than the closely related chimpanzee.

5.1.2 Social life

The basic social organization of non-human primates varies considerably during the year, due to food accessibility, female reproductive state and birth period. There is also variation between different species: from solitary orangutans to sociable chimpanzees and bonobos. The orangutangs are solitary animals that usually live in mother-child groups, while gibbons form groups that resemble human families, with a mated pair with their offspring. Gorillas, on the other hand, live in somewhat larger stable groups, usually centred around one dominant male, the silverback male. A gorilla group may consist of one silverback, one younger male, three or four females, and about five youngsters, under eight years of age (Fossey 1983, Goodall 2005). Chimpanzees and bonobos have more flexible groups, the so-called *fission-fusion* social unit mentioned earlier (Kummer 1971). This means that their size and composition not only change according to season of the year, but also during the day. Typically, chimpanzees disperse in smaller subgroups for short periods, and then meet again. The reestablishment of relationships is shown in ritual greetings and procedures of conciliation, expressed both by vocalizations and by tactile signals such as hugging, kissing, and grooming. In these small groups, consisting of around five individuals, they travel in the trees to forage and rest. There are different labels suggested for the different groups; Goodall (1968) uses the term *community*, for a group of 50 to 100 individuals, and *party* for a smaller group. The members of a party group have visual contact with each other and auditory contact with the others in the community. Party sizes range between 10–20% of the community, and the time the party members spend together ranges from some minutes to a couple of hours (Goodall 1986, Boesch & Boesch-Achermann 2000). Most party groups seem to be foraging, but there are also reports on subgroups of adolescent males who join together to groom (Mitani & Amsler 2003). DeVore (1968: 20) assumed an essentially life-long membership; "an individual is born, matures, leads its adult life, and dies in the same group", but later field research has found that while this may be true for males, females emigrate from their birth community to join another group (Nakamura & Itoh 2005).

The social life of bonobos is quite similar to social life in chimpanzees, with a fission-fusion character, and females emigrating to new communities. One difference, however, is that bonobos have a more egalitarian social structure, with the females holding a stronger position than the chimpanzee females. When emigrating, the bonobo females integrate in the new community by bonding with senior female members of the new group. The alliance lasts until the emigrated female has had offspring (Kano 1982, 1992, de Waal & Lanting 1997).

Captive nonhuman primates live under different conditions, where neither the flexible fission-fusion life nor the emigration is an option, but humans decide which individuals stay together in a group, and for how long. As studies of captive chimpanzees have shown, there is a large potential for adjustments in the social behaviour of the species. This opens up for another dimension in research of nonhuman primates – the ecology of communicative behaviour. If there are different communicative behaviours in different environments, this suggests plasticity in the communicative system (Bard 1998, Leavens et al. 2005). For example, captive chimpanzees have been found to be able to accommodate to a situation of reduced space and permanent groups, by increasing the amount of appeasing behaviours, such as grooming and submissive greetings (de Waal 1994). Even if the aggressive behaviour also increases, the effect of aggression is minimized by the appeasement signals. On the basis of these findings, de Waal (1994: 248) proposes a *coping model*, where "spatial crowding results in an increased risk of aggression, to which the apes react with calming gestures that serve to reduce this risk".

Mother-offspring interaction differs between the different species. Typically, the primates belonging to the Great Apes carry their infants for the first years of their lives. The chimpanzee infant, for example, is in more or less constant body contact with the mother during the first months and it is difficult to observe any other communicative behaviour than this tactile contact. In the beginning, the baby is not able to hold on the mother's fur, but the mother gives support by her hand. After a couple of weeks the infant is able to cling to the mother's fur and then ride on her back. The close mother-infant body contact comes to an end around the age of 8 months, when the infant starts to make short excursions from the mother. Vocalizations occur, for example when the mother and infant wish to re-establish physical contact.

To find nonhuman primates with cooperative breeding, where others than mothers assist in caring for the young, we have to leave our closest relatives and study New World Monkeys. Here we find cooperative breeding in the *Callitrichidae* family (Hrdy 2009, Snowdon & Pickhard 1999). This family includes for example marmosets and tamarins. There is more caretaker-offspring communication in this family than in chimpanzees, involving not only the biological parents but also other adults in the group. In a cooperative breeding context the need for acoustic and visual communication is different than for the situations when the infant is in constant close body contact with the mother.

5.1.3 Studying nonhuman primates – how it all began

Much of what we know about social life in wild chimpanzees stems from fieldwork from six key sites in Africa, where projects were initiated in the 1960s and 1970s. In 1960 Jane Goodall started her pioneering work in Gombe National Park in Tanzania (Goodall 1965, 1967, 1986). The results from Gombe, where Goodall has been carrying out research for more than 40 years, have had a great impact on our understanding of the social life of nonhuman primates. A few years after Goodall, another study set off in the same region, this time in Mahale Mountains (Nishida 1968). Then two projects were started in Uganda, one in Budongo (Suzuki 1969) and one in Kibale (Ghiglieri 1984). Kortlandt (1962) investigated chimpanzees in Bossou in Guinea and Boesch & Boesch-Achermann (2000) reported from a longitudinal study, starting in 1979 in Taï Forest of the Ivory Coast.

Before these field studies we knew very little about primate behaviour outside captivity. One of the very first efforts to find out more was the experiment by Richard Lynch Garner, who in 1868 placed himself in a cage in the forest to be able to observe chimpanzees without exposing himself to danger (Reynolds & Reynolds 1965: 394). Apart from Garner's experience in 1868, the earliest investigations of behaviour in nonhuman primates took place in laboratories and in zoos, where isolated animals were studied (e.g. Darwin 1872/1965, Bingham 1927; in Bingham's case it was a chimpanzee mother with her child). DeVore (1973) compared these early studies to what would be the result if we built our knowledge of human behaviour from "a study of inmates in a maximum-security prison". However, the environment in zoos has changed over time, and the captive nonhuman primates today find themselves in conditions that are far more similar to natural conditions than they used to be, with groups of animals, and in enclosures enriched with vegetation etc.

Studies of nonhuman primate communication that are undertaken today can be placed into three categories: studies of social groups of primates in the wild, studies of groups in captivity, and studies of individuals taking part in teaching experiments.

5.1.4 Observation techniques

Research on nonhuman primate communication can be carried out in wild populations or with captive individuals; it may be observational or experimental, long-term or short-term. Studying chimpanzees in the wild is a demanding work, to say the least. Since the animals spend much time foraging high up in trees and also travel at tree-level it is difficult to film them, so the researcher has to follow by foot and observe from the ground. Furthermore, the animals have to be habituated to humans in order not to react by fear. This means that the researchers have to wait for a long time before they can start making detailed observations. For example, Jane Goodall reports that she had to keep at a distance for the first 18 months before she was able to come close to her subjects. In the study by Boesch and Boesch-Achermann (2000) it took about two years before the chimpanzees were used to humans being around and behaved naturally. According to the Observer's paradox (Labov 1972), even if the observer's intention is to study undisturbed, natural behaviour, the very fact that an observer is there will have an effect on the behaviour – and make it less natural.

In the wild, data are usually collected by observations, recordings and playback experiments. Observation of communicative patterns may be difficult since it is impossible to notice everything that goes on in dynamic interaction, where individuals influence each other's behaviour and adjust to the others in the group. Two observation methods that are often used are *scan sampling*, where a whole group is observed at regular periods of time, their behaviour noted at each period, and *focal animal* sampling, where the researcher focuses on one single animal at a time (Altmann 1974). To obtain reliable data, observations are often combined with some kind of recordings. Audio recordings give possibilities for detailed analyses by different techniques, such as bandpass filters. Video recordings can be used for detailed analysis of postures and movements by frame-by-frame analysis. In order to follow animals at some distance, a global positioning satellite can be used to follow the subjects' movements (Arnold & Zuberbühler 2006).

The development of the playback technique was a crucial step towards controlled studies on vocalizations and their associated meanings. By taking the recorded vocalization from an ape, and playing it out in the wild through a loudspeaker, the possibility of documenting the reactions from conspecifics increases substantially. However, playback experiments have the problem that they may influence the behaviour of the subjects and it is therefore important not to do it too often. Seyfarth et al. (1980b) suggest restrictions that should be effected, for example always to have at least 24 hours intervals between each playback.

Studying animals in captivity is different from observations in the wild. While there is a certain disadvantage for the researcher on captive animals to claim natural behaviour in the social life of captive apes, there are also positive factors, such as clearer views, opportunity of constant surveillance, individual recognition, possibilities of setting up experiments and thereby formulate and test research questions that are different from those found in descriptive observations. For example, studies on gaze and joint attention (e.g. Bard et al. 2005, Call et al. 1998, Leavens et al. 1996, Tomasello et al. 1998) are based on animals in captivity.

Another important methodological issue is how to describe and categorize animal communicative expressions. For vocalizations, different methods have been in use, for example some kind of transcriptions (calls from vervet monkeys may sound like *nyow* and *rraup*; Struhsaker 1967: chimpanzees say *wraa* and *huu*; Goodall 1986). There is no consensus as to how acoustic signals in nonhuman primates should be described and there is a huge variation between research groups. Sometimes the vocabulary for human vocalizations is used (e.g. screams), other methods are to illustrate sounds by paralinguistic transcriptions, based on onomatopoetic principles or use articulatory or acoustic descriptions. The box below summarizes some different ways to describe nonhuman primate vocalizations.

Nonhuman vocalizations have other qualities than human sounds and it is difficult to give an adequate illustration by using the same type of descriptions. Probably the best way of illustrating nonhuman primate vocalizations would be to put the recordings on the internet.

Box 5.1 Descriptions of vocalizations of nonhuman primates

1. Terms used to describe human vocalizations: *scream, call, laughter* (e.g. Goodall 1986: 127)
2. Use of paralinguistic transcriptions: *waa, waaoo, aaoo* (Crockford & Boesch 2003: 117), *woof-waa, rraugh, rraup* (Struhsaker 1967: 313), *boom* (Zuberbühler 2002: 293)
3. Description of the articulatory processes behind the production: *exhalation and inhalation* (e.g. Goodall 1986)
4. Terminology referring to acoustic analysis: *duration, fundamental frequency* (e.g. Riede et al. 2004, Crockford & Boesch 2005), *broad-band (noisy) calls* (Digweed et al. 2005)

5.2 Communicative functions

When discussing communicative functions in human interaction we have the advantage that we are involved and can look at what happens both from the inside and from the outside. We have experience from earlier situations in life, and can easily conclude when the utterance "How are you" is meant as a request for referential information (if a doctor asks) or a social geeting formula (when meeting a colleague at work such social formulas are important – not saying anything is perceived as rudeness). Thus, we contextualize the activity to understand the function. When it comes to communicative functions in other species it is more difficult to define the reason behind the interaction.

In the literature, three ways are discussed to understand the functions: (1) to apply an ecological perspective and start from what is needed in the environmental context. An example of this is Collias' (1960) suggestion that the major functions of animal interaction have to do with food, enemies, reproduction and group movements; (2) to use the situation as a point of reference. For example, Struhsaker (1967) was able to link situations such as "snake predator" to specific signals from vervet monkeys.; and (3) to connect specific expressions to emotional states. By this method, Goodall (1986: 127) listed 13 emotions or feelings to be associated with different chimpanzee calls.

In this chapter we will use the distinction between the referential and the social function as our point of departure. As will be clear from the following, it is the search for the referential function that has instigated the main bulk of research. The referential function gives specific information about something external (e.g. a predator) and has often been seen as the hallmark of human language.

5.2.1 The referential function – first reported in vervet monkeys

As mentioned in the introduction, the first report of a referential function in nonhuman primate communication in the wild was Struhsaker's (1967) study of vervet monkeys (*Cercopithecus aethiops johnstoni*) in Kenya. Struhsaker observed that vervet monkeys use different alarm calls to warn for different classes of predators, of the types eagle, snake or leopard. The first step was to identify a relation between a vocalization and a particular event. Next step was to use the playback technique and test whether it was the particular vocalization, or the sound in combination with other signals, that led to a specific effect. In later studies (e.g. Seyfarth et al. 1980a, b, Cheney & Seyfarth 1982, Züberbühler et al. 1997, Arnold & Züberbühler 2006) the playback method could confirm the original observations on vervet monkeys, as well as extend the findings to other nonhuman primates. In Box 5.2 an overview of some of the studies on referential calls in nonhuman primates is given.

Box 5.2 Examples of studies investigating referential communication

COMMUNICATIVE FUNCTION	SPECIES	METHOD	MAIN FINDING	REF.
Various	Vervet monkey Kenya	Observations, audio recordings	A catalogue of calls with certain functions, e.g. warning for different predators	1
Alarm call	Vervet monkey Kenya	Observations, playback, filming	Evidence for semantic signalling, three different alarm calls	2
Alarm call	Diana monkey Cote d'Ivoire	Playback	Semantic calls (eagle, leopard), different alarm calls in males and females	3
Alarm call	Diana monkey & Campbell monkey Cote d'Ivoire	Playback	Diana monkeys understand Campbell monkeys	4
Alarm call	Diana Monkey Cote d'Ivoire	Playback	Semantic signals, with syntactic modification	5
Alarm call	Putty-nosed monkey Kwana	Playback	Combinations of two basic alarm call types give new meaning	6
Alarm call	Chimpanzees Cote d'Ivoire	Observation (1 044 hours)	Two types of calls, hunt/snake	7
Alarm call	White-faced capuchin Costa Rica	Recordings (113 bouts of calls)	Two variants of calls: one for birds and one for other predators	8
Alarm & contact	Chacma baboon Botswana	Recordings (contact and alarm bark). Playback	No difference in response, no running to trees at alarm call	9
Travel direction	Chimpanzee Cote d'Ivoire	Observation	Drumming in different combinations used to signal for direction and resting	10
Food call	Chimpanzees Uganda	Observation	Pant-hoots at fruiting trees indicate high status males	11
Food call	Tufted capuchin monkey Argentina	Recordings (96 hours)	Food-associated calls functionally referential, audience effect	12
Social info	Vervet monkeys Kenya	Playback	Five grunt types, "rudimentary" referential	13

References: 1. Struhsaker 1967 2. Seyfarth et al. 1980b 3. Zuberbühler et al. 1997 4. Zuberbühler 2000b 5. Zuberbühler 2002 6. Arnold & Zuberbühler 2006 7. Crockford & Boesch 2003 8. Digweed, Fedigan & Redall 2005 9. Fisher, Metz, Cheney & Seyfarth 2001 10. Boesch & Boesch-Achermann 2000 11. Clark & Wrangham 1994 12. Di Bitetti 2005 13. Cheney & Seyfarth 1982.

The referential function is not only found in acoustic signals. As discussed in Chapter 4, *mutual gaze* (eye-to-eye contact) is assumed to be important for human mother-infant bonding and particularly for the development of language and cognition (Bowlby 1973, Berko Gleason 1977). From the earliest period of the infant's life, mothers follow their gaze and comment on what they are looking at. This is assumed to encourage children following the gaze of others, which in turn is a sign of representational ability. For a long time, mutual gaze was regarded as a typical human activity and not expected to occur at all in nonhuman primates. However, studies on nonhuman primates have shown that they also use mutual gaze (Bard et al. 2005), as well as gaze following (Tomonaga et al. 2004) and gaze alternation (Schino et al. 2006). Furthermore, hand or arm pointing, another characteristic of referential communication, has also been found to occur in nonhuman primates (Menzel 1999, Leavens et al. 2005, Veá & Sabater-Pi 1998).

Warning calls (also called alarm calls) are the most studied nonhuman primate vocalizations to date. As mentioned above, Struhsaker (1967) presented the first detailed description of different vocalizations in vervet monkeys. He identified a system of distinct alarm calls used for different predators; i.e. with referential function. As was summarized in Box 5.2, later studies have shown that also other nonhuman primates use differentiated alarm calls.

The *food call* is another example of a call referring to information outside the animal's own state of mind. Its communicative function is not yet clear, as it is not automatically uttered as soon as an individual finds food, but certain conditions have to be fulfilled. For example, for chimpanzees it has been reported that they are more likely to call when approached by higher-ranking individuals (Clark & Wrangham 1994), whereas white-faced capuchin monkeys (*Cebus capucinus*) are shown to call more with fruit than with insects or eggs (Gros-Louis 2004), tufted capuchin monkeys (*Cebus apella nigritus*) seem to be sensitive to the distance from the audience – if the other individuals were close the call came earlier than when others were at a distance (Di Bitetti 2005) and finally, rhesus monkeys call more when they are hungry (Hauser & Marler 1993a, b). Further research is needed before we can tell whether it is the case that food calls are differently distributed in different species, or if the same factors trigger them across different nonhuman primates. Their function is puzzling. As in alarm calls, there is a subtle balance between the cost and benefits in calling. To have a behaviour that loudly announces an enemy can be fatal if it means discovery by a predator, but it is beneficial to the group. Similarly, the food call may result in the individual losing the food to someone else, but on the other hand cooperation is advantageous to the group as a whole. Thus, the food call may well be a candidate for *altruistic* behaviour, since it implies giving up something in order to help another individual finding food (Hauser 1996). On the other hand, it may also be a selfish way to make sure there are enough individuals around to help in predator surveillance (Bradbury & Vehrencamp 1998).

The finding that nonhuman primates use referential functions has caused a lot of discussion, since referential function is one of the features on every list on what is unique in human language. It implies that nonhuman primates would have the potential for symbolic communication, which has been suggested as the hallmark of human language. The question is if a referential function in animal communication means the same as in human communication. If nonhuman primates react appropriately, it seems as if they all share the same meaning of an arbitrary signal. But what do the vocalizations really mean? Does the monkey use the referential function and inform the others that "an eagle is coming", or is it a matter of an automatic reflex, to call when seeing a predator (see Zuberbühler et al. 1997: 603)?

5.2.2 The social function – with focus on synchronization of behaviour

The social communicative function of nonhuman primates have not attracted the same interest from researchers as the referential function. One reason may be the methodological problems. In humans the social function is discussed as something that has to do with cooperation, mutual understanding and group solidarity. This is difficult to identify in species outside our own, compared to the relative ease of setting up a playback experiment.

Synchronization of behaviour is one feature characterizing the social function in humans, found both in verbal and nonverbal communication. Studies investigating synchronization in nonhuman primates, have found this behaviour to occur also in gibbons (Geissmann 2000, 2002), chimpanzees (Mitani 1994, Mitani & Gros-Louis 1998), Japanese macaques (Suguira 1998) and Campbell monkeys (Lemasson et al. 2003, Lemasson 2011).

The most striking examples come from the gibbons (genus *Hylobates*) Their vocalizations are so distinct that the different species are identified by vocalization patterns rather than their body structure. A particular vocalization that can be heard in the rainforests of Asia, is the gibbon song, described in the passage below:

> I have heard the calls of wild gibbons ringing through the forest in north-eastern India, and the first time I heard them, they sounded so human that I wondered, just for a moment, if they came from high-spirited young people. (Burling 2005: 125)

The *duet song* is performed by the pair of gibbons, with different parts for the female and the male. It begins with short phrases from the male and then there is a female 'great-call'. During the build-up phase of the great-call, the male is silent, but at the climax he joins again (Geissmann 2002). Typically, the songs are performed in the early mornings and males may sing solo songs even before sunrise. The function of the duet song has been suggested to be the strengthening of pair bonds. This seems reasonable, since gibbons live in long-term and monogamous relationships and duet singing correlates with mutual grooming, behavioural synchronization and distance between mates, other features that are known to strengthen bonding (Geissmann 2000) – but whether the duet singing is the cause or the effect of the strong pair bond is still unclear (Mithen 2005). At the same time as it strengthens the bond, the song may function as a spacing mechanism, to tell others that the territory is taken.

Other studies have analyzed synchronization on group level, and the results show that there is vocal sharing in stable groups. In a study on chimpanzees, a relation was found between grooming preferences and vocal synchronization, so that those who groomed each other also are those doing most of the vocal sharing (Mitani & Gros-Louis 1998:1042). In Japanese macaques, Suguira (1998) reports that vocalizations have both stable features, presumed to code individuality, and flexible features, matching the preceding call. Groups with close affiliations are the ones with most vocal synchronization (Suguira 1998). Campbell's monkeys synchronize vocalizations in a way that "Level of vocal sharing is correlated to level of social integration" Lemasson (2011:53).

5.3 Acoustic communication

Above, functions of primate communication were presented. Now we will take another perspective and use the sense as point of departure. For nonhuman primates, there are four main sensory modalities, or main channels for communication, the visual, the acoustic, the tactile and the chemical (or olfactory) channel. In real situations, communication is effected through two or more channels simultaneously, but in this presentation we will keep the channels apart and discuss them one at a time.

Nonhuman primates use a whole range of behaviours for acoustic communication. Some sounds are vocalizations, such as calls, grunts, pant-hoots, barks, purrings, growls; other sounds are produced by lip-smacking, teeth-chattering, breast-beating, drumming on trees, breaking branches, even jumping on dry sticks. In this section, the main focus will be on vocal communication. There is a large variation among nonhuman primates as for how vocalizations are used, due to many different factors, for example breeding traditions (cooperative breeders tend to vocalize more) and ecological conditions (animals living in dense forests vocalize more). We will start by describing the chimpanzee, and then give some examples from research on other nonhuman primates, such as macaques, gibbons, and marmosets.

5.3.1 Some vocalizations and their use

Goodall (1986) uses the method of linking sounds to emotions and identifies more than thirty different vocalizations among the chimpanzees of Gombe. A particularly significant vocalization is the so-called *pant-hoot*, described by Goodall in the following way:

> Pant-hoots, voiced both on exhalation and inhalation, are the most commonly heard call of adult individuals and, more than any other (at least to human ears), serve to identify the caller. (Goodall 1986:134)

The pant-hoot serves as an identity marker for the individual, and it seems to function also as a group-marker, in a way similar to human dialects. In an investigation of long distance pant-hoots in two chimpanzee populations, Gombe and Mahale, in Tanzania, acoustic differences were found, with chimpanzees in Mahale using shorter and higher pitched elements, than the chimpanzees in Gombe (Mitani et al. 1992). It has been speculated whether group differences are caused by genetic variation, differences in body size, environment, and/or habitat. Another explanation may be social learning, that the young chimpanzees "converge in subtle features of pronunciation" (Mitani 1994:206) and thus accommodate their calls to the calls of adults. This hypothesis of vocal learning has been supported by studies comparing DNA and habitat between groups (Crockford et al. 2004). Some chimpanzee vocalizations have been linked to an individual's emotions, such as fear (*Wraaa*) or puzzlement (*Huu*), whereas others have been interpreted as referential, giving information about food and danger. We discussed referential aspects above, and here we will focus on social aspects of vocalizations.

Contact calls have not been subjected to the same systematic examination as alarm calls and food calls. One problem for the researcher is that contact calls do not generate immediate response from conspecifics, but it seems as if the individual may choose whether to react or not. A key factor role seems to be kin relationships, since the playback of a call from a kin produces more interest than a signal from a non-kin. This suggests an ability to categorize individuals on the basis of kinship – and to put that information in the features of the call. In a study with playback of contact calls (or 'coos'), Rendall et al. (1996) found that rhesus macaques (*Macaca mulatta*) could discriminate between matrilineal relatives and other group members. The acoustic analyses of the calls indicate certain structural similarities, especially among related females, implying that features of the contact calls are learned though a process of convergence (Hauser 1996, Rendall et al. 1998). Rhesus macaques live in multi-male and multi-female groups of between 6 and 90 members, with three times more females than males (Makwana 1978). The groups are organized around closely related females. Males emigrate when they are sexually mature, whereas females stay in the group.

The bonnet macaque (*Macaca radiata*) is known for its rich repertoire of vocalizations. Hohmann (1989) identified 25 basic patterns some of which varied according to the situation and the age of the animal. For example, some calls were used exclusively by very young infants, whereas other were typical for older infants and juveniles. There seems to be a learning component in the vocalizations and juveniles are found not be as proficient as adults to discriminate between different calls. Since infants spend their first months clinging to their mothers they probably learn to associate alarm calls with fleeing by the reactions from their mothers (Ramakrishnan & Coss 2000).

The gibbon song was presented above as an example of vocal synchronization. The fact that this song is similar in captive and wild gibbons suggests that it is innate (Geissman 2002), but the picture is more complex. Individual differences suggest that there are dialects (Cheyne et al. 2007) and also differences over time suggest that there is a learning component. In a detailed longitudinal study of a young female gibbon (*Hylobates gabriellae*) participating in her mother's great call, the song was found to develop gradually in phases between the ages of 5 and 36 months (Merker & Cox 1999). The early parts of the call developed first, whereas the final parts were not complete even after 2 years of practice. Interestingly, Geissmann (1983) gives an example of a widowed female singing both parts of the duet, which shows that females have the capacity for both the female and the male song.

5.3.2 Vocal learning in nonhuman primates

A lot of interest has been directed towards the development of vocal behaviour of the *Callitrichidae* family, belonging to the New World primates. Pistorio et al. (2006) made a careful acoustic analysis of the vocal development of the South American arboreal species common marmoset (*Callithrix jacchus*) from the age of 3 weeks up to the age of 25 weeks. The analyses demonstrated significant changes over time, as the early vocalizations disappeared and the calls became more and more adult-like. Another example is the work on the pygmy marmosets (*Cebuella pygmaea*) where ten different vocalizations have been described (Pola & Snowdon 1975). The vocalizations are used to communicate threats, fear, rage, submission, and social contact. A closer study of these contact calls showed that they are plastic and undergo significant changes during infancy (Elowson & Snowdon 1994, Elowson et al. 1998, Snowdon & Elowson 2001). The change correlates with parental dependence and age of infant. The infants start vocalizing in their first week of life, producing repetitive and rhythmic sounds that resemble babbling in human children. This early babbling functions as vocal practice, and infants that babble more frequently are found to have better formed vocalizations at 5 months, than other infants (Snowdon & de la Torre 2002). Just like human infants, the marmoset infants produce vocalizations that are a subset of adult production, but seemingly without a referential meaning. Just like human caretakers, the marmoset caretakers react by vocalizing back, approaching or picking up the infant. The *Callitrichidae* family has been pointed out as the prime example of cooperative breeding and the marmosets are no exception. Their social group consists of parents, older siblings and often also some unrelated individuals, and all group members take part in protecting, carrying and grooming the infants. The family setting and spontaneous vocalizations in the marmoset infants make this group interesting in the comparison nonhuman – human communication: "..social parallels make the Callitrichids a more compelling analogous group to study than the phylogenetically closer, but socially dissimilar apes (Elowson et al. 1998: 32).

5.3.3 Structural aspects of vocalizations – do nonhuman primates have syntax?

One of the assumed differences between human and nonhuman vocal communication is that human speech is built upon a system of discrete categories, which can be combined in different ways, whereas the expressions in nonhuman primate communication are described as gradual. A number of experimental studies have been undertaken to investigate whether rudiments of human language may be found also in the vocalizations of nonhuman primates (e.g. Arcadi 1996, 2000, Fischer et al. 2001, Zuberbühler 2002). In one of these experiments, Zuberbühler (2002) identified a sound (described as a low resounding call, a 'boom') that could be combined with another sound and change its original meaning. A boom indicated that whatever message followed did not have the usual meaning, but a slightly different, similarly to modifying words like "a kind of" in human language. To give an example, when the Campbell monkeys (*Cercopithecus campbelli*) give an alarm call indicating an eagle, the Diana monkeys (*C. Diana*) respond by giving their own eagle alarm call, but when there is a boom preceding the alarm call, they do not respond in the same way. The meaning of the call has changed from a precise predator label, into a signal of general irritation. This suggests a potential of syntax, in the sense that two elements can be combined to form new meanings. If this turns out to be general property in nonhuman primate communication, yet another suggested difference between human language and communication in nonhuman primates has been found to be quantitative instead of qualitative. Zuberbühler (2002: 298) suggests that in teaching language to nonhuman primates what researchers are doing is to tap into "a species' naturally existing syntactic abilities".

5.4 Visual communication

Nonhuman primates basically have the same body structure as humans, and the posture can give information about an individual's emotional status in a similar way. Body posture and movements, facial expressions, hand movements and hair bristling are used in visual communication.

5.4.1 Body postures

Strength is usually signalled by an erect position, and may be further reinforced by swinging the arms, marching forcefully and hurling rocks and branches. Nonhuman primates also have a possibility to display aggression or excitement by erecting the body hair, particularly the hair on the back and shoulders. Goodall (1986) reports that the hair alternates between rising and falling, according to the emotional status of the moment. Subordinate individuals react by moving away when a superior shows bristling. Submissive chimpanzees have sleeked hair and a stooped position, trying to look smaller than they are.

5.4.2 Hand and arm gestures

Which movements that constitute communicative gestures is difficult to define in humans, and also in nonhumans. A number of different definitions occur in the literature. As an operational definition of gesture Scott and Pika (2012) suggest the following: "any nonvocal bodily action directed to a recepient that is mechanically ineffective and represents a meaning, beyond itself, that is manifested by others of the social group" (Scott & Pika 2012:158–159). This definition can be applied to the analysis of gestures in humans as well as nonhumans. However, although this definition covers any "bodily" action, most studies of gestures have focussed on movements of hands and arms.

Hand and arm gestures have been studied in a number of nonhuman primates, for example bonobos (e.g. Pika et al. 2005), chimpanzees (e.g. Goodall 1986, Bard et al. 2005, Leavens et al. 2005), gorillas (e.g. Pika et al. 2003), macaques (e.g. Schino et al. 2006) and mandrills (Laidre 2012). Some hand gestures are observed so frequently that they have been given a label, for example the chimpanzee "arm raise" (to solicit grooming, or to appease someone; Plooij 1979), the gorilla "arm shake" (Pika et al. 2003), and the mandrill "hand extension" (to provoke someone; Laidre 2012). Other gestures have gained reputation because of the cultural learning, for example the "grooming hand clasp" (de Waal 2001).

One of the most studied hand gestures is pointing. Hand pointing and gaze following share one property, namely referring to some outside object or action. Just like gaze following, pointing has been suggested by some researchers to be unique to humans ("… pointing separates humans from primates, just like the use of language does" Kita 2003:2). However, other researchers claim that pointing occurs also in nonhuman primates (e.g. Leavens & Hopkins 1999). Menzel (1999) reports on spontaneous pointing occurring in an experiment with the language-trained female chimpanzee Panzee. At the time of the experiment she knew 256 words, or lexigrams. The experiment was aimed at investigating Panzee's memory abilities, but an unexpected result was that Panzee was able to use pointing as a useful means to get what she wanted. The investigator hid food and other objects in an outdoor confinement, visible to Panzee indoors. The next day Panzee was able to instruct an uninformed person, where to find the desired object. The assisting person did not know that anything was hidden, but Panzee could perform correct directions by a combination of communicative signals: pointing outdoors with her arm, making the sign for "hidden" by covering her eyes with her hand, and pointing at the screen for the lexigram that corresponded to the object. This experiment shows that chimpanzees are able to use pointing in a similar, if not exactly the same way, as a human would have done it in the same situation.

The controversy about pointing as uniquely human, or also used by nonhuman primates, is partly due to different definitions of pointing, and distinguishing pointing from reaching behaviour. One important issue is the use of the index finger. In many human cultures the index finger pointing has become the prototypical way of pointing (this is illustrated by the fact that many languages, like English, use the term index or pointing for the finger next to the thumb cf. Chapter 4). However, even if the index finger pointing is the most common in Western cultures, research from other parts of the world have reported that humans may also point with the whole hand, body or head turn, eyes, or mouth. This behaviour is more similar to pointing in nonhuman primates, who also tend to use the arm, the whole hand, and the body and/or head direction combined with eye gaze. Another issue under discussion is the fact that pointing occurs more frequently in captive chimpanzees than in chimpanzees in the wild, even if the captive chimpanzees have not been subjected to specific training. The reason may be human influence, but it may also be due to the specific constraints of the captive situation. The main function of pointing among captive chimpanzees seems to be connected with getting to something that is out of reach, e.g. food, rather than to direct the attention of others to some object. Leavens et al. (2005) suggest that a pointing gesture is trigged by precisely those situations where an individual wants to reach something but is hindered by a confinement (e.g. chimpanzees) or by immobility (e.g. human infants). In young human children the function of directing other's attention develops around the age of one year, and then pointing is typically used together with vocalizations, e.g. in order to name objects (e.g Goldin-Meadow & Butcher 2003).

Some issues in the discussion on pointing in nonhuman primates:
- With index finger?
- Together with vocalizations?
- Together with eye?
- Mostly in captivity?

The reason why nonhumans do not use poining regularly is probably that they find no meaning in it. In experiments where food is placed under one of several containers, and a human experimenter points at the container, the chimpanzees can follow the direction of the point, but then they take an arbitrary container. "When following the human's point with their gaze, all they perceive is a useless bucket" (Tomasello & Moll 2010: 340). It is a different matter when the experimenter reaches for the container – then the ape competes with the human to reach the goal (Hare & Tomasello 2004).

5.4.3 Face and gaze

The facial muscle controls in nonhuman primates are similar to those in humans. The mouth region is particularly important, where baring the teeth may be interpreted either as a threat signal or a submissive signal, depending on whether the corners of the mouth are drawn down or lifted up (cf. the 'frequency code' mentioned above). Of other facial expressions, a pout for distress, a play-face expressing good humour, and various lip movements in relaxed situations, are commonly mentioned in the literature. Just like in humans, most facial expressions are used in combination with movements and vocalizations.

The eyes and gaze serve as strong signals with quite different functions. A direct gaze can be interpreted as a threatening stare. In spite of this, it also has a completely opposite meaning, namely that of social bonding. Mutual gaze is much used in caretaker-infant interactions, and has been assumed to play an important role for communicative development in humans. However, this behaviour is also found in nonhuman primates. Bard et al. (2005) compared mother-infant interaction in chimpanzees and humans and found the amount and length of mutual gaze to be relatively similar. Furthermore, they found differences between groups of chimpanzees, suggesting that a certain amount of social learning is taking place. In some chimpanzee groups the mothers and infants spend more time looking at each other, whereas in other groups less time is spent on mutual gaze, and more on cradling, grooming and bodily contact between mother and infant. This inverse relationship between mutual gaze and bodily contact can also be found in human societies (e.g. Crago 1992).

In addition to the mutual gaze, also the direction of the gaze has an important function. In human groups, looking in a certain direction functions in the same way as pointing, and there is empirical evidence that this is possible also in other animals. In a natural environment it is problematic to assess the role of gaze in communication, for example if several signs are given simultaneously, or the subject was heading in the target direction for quite other reasons, but by setting up experiments it has been possible to control for confounding factors. Tomonaga et al. (2004) investigated the development of gaze following and found young chimpanzees to be able to follow pointing cues at the age of 9 months, head-turn cues at 10 months, and eye-gaze cue at 13 months. Tomanaga and colleagues (2004: 232) conclude that "gaze following abilities in chimpanzee infants seem to develop gradually and in a step-by-step manner, as has also been found in human infants". However, gaze-following abilities is not restricted to chimpanzees, but also found in other nonhuman primates. Tomasello et al. (1998) investigated gaze following in five different primate species: chimpanzees, sooty mangabeys (*Cercocebus atys torquatus*), rhesus macaques (*Macaca mulatta*), stumptail macaques (*M. arctoides*) and pigtail macaques (*M. nemestrina*). The method was to have a coordinator standing in an observation tower and showing an object to one selected individual while observing what the others were doing. This made it possible to factor out the possibility that the subjects were looking by accident. The results showed that there were no difference between different species but gaze direction was used as a cue to external referents in more than 80% of the cases. Meunier et al. (2008) found that high-ranking white-faced capuchins (*Cebus capucinus*) monitored the movements of the group by looking backwards and waiting for the other to follow.

A question that has been discussed extensively is whether nonhuman primates really intend to share knowledge, for example by gaze, or it is something that has more to do with the receiver being a meticulous observer. In a study on Japanese macaques (*M. fuscata*) the results suggested that communication by gaze could be intentional (Schino et al. 2006). The macaques, like many other primates, form coalitions with each other and help each other in conflicts. Schino et al. observed that the macaques asked for assistance in aggressive situations by turning the head and look alternatively at the opponent and a potential helper. Not anyone could be a candidate as a helper, but there was a clear avoidance to signal to anybody related to the opponent when asking for help. The interpretation is that the macaque is aware of the danger of asking someone from the "wrong" family, knowing that this individual would rather support his own kin instead. The recognition of family relationships seems to be an important factor in the social life of nonhuman primates, and chimpanzees are even able to recognize kin relationships on the basis of facial features on photographs (Parr & de Waal 1999).

5.5 Tactile communication – a lot of grooming

The grooming behaviour in a group on nonhuman primates is as typical as human chatting and talking, and it is assumed to have the same function – reduce aggression and serve as oil in the social wheel (Dunbar 1991, 1996). Grooming can also be regarded as an investment for the future, since social bonds and alliances are formed in grooming networks. At the same time, grooming has a practical function for cleaning the body from unwanted particles. It is often directed to parts of the body, which the individuals are not able to reach themselves.

Most observations of chimpanzee groups report grooming to be most frequent between males, and directed towards a male of higher rank. Males that are close in rank tend to groom each other reciprocally more than they groom other individuals (Arnold & Whiten 2003). The female chimpanzees often spend all time with the infants and are not involved in adult-adult grooming. However, there are also situations when female chimpanzees take parts in grooming. In an account of how a newly immigrated chimpanzee female settles down in a new group, Nakamura & Itoh (2005) describe a substantial amount of tactile communication. During the hours of observation, this female copulated with four males, and was groomed by seven individuals, three females and four males. She only groomed back one of the others, one of the adolescent male that she had copulated with. Grooming is related to status, and since the chimpanzee social group is male-dominated the majority of grooming occurs between the males. In species where females have higher status, like the bonobos, more grooming between females could be expected. Studies have given contradictory results however. Some studies report that it is the high-ranking females that receive most grooming (e.g. Franz 1999), whereas other studies show that it is the bonobo males (Stevens et al. 2006). These discrepancies in results reveal that there are probably a lot of local differences.

Close analyses of structural aspects of grooming practices reveals some small, but systematic, differences also between chimpanzee groups. For example, a hand-clasp grooming, with the participants holding hands in the air, is typical to chimpanzees in Mahele in Tanzania, but does not occur at all in Gombe, only 170 kilometers away (de Waal 2001). This hand-clasp grooming is not used by all individuals, but is probably socially constrained to occur only between adults, since adolescents' attempts to initiate hand-clasps are ignored (Nakamura 2002). Other local features are a particular kind of social scratching with erect fingers, and a sputting sound that goes with the grooming, both found only in Kibale in Uganda (Nishida et al. 2004).

Some of the variation found in tactile communication in different groups has ecological explanations. For example, gorillas staying in the cold outdoors have more physical contact than gorillas spending their time indoors (Maestripieri et al. 2002). In other instances the reason for differences between local groups may be attributed to cultural learning (see below).

5.6 Chemical communication

Olfaction has been called the "neglected sense" in primate behaviour (Heymann 2006) and not much is known about its function and structure, partly due to lack of reliable observation techniques. In many species, chemical signals are important to regulate territorial claims, personal relationships and inter-sexual communication, and it is plausible to assume the same function for nonhuman primates. However, the empirical evidence is still to be found. Some of the few observations are reported by Goodall (1986), who describes how both male and female chimpanzees in Gombe inspect each other's genital area by touching with the fingers and then sniff them.

5.7 Cultural/dialectal differences – results of social learning

Most studies on culturally, or socially, learned behaviours, in nonhuman primates have dealt with how they use material to build nests or find and eat food (e.g. Goodall 1986, Boesch & Boesch-Achermann 2000, McGrew 2004). When it comes to cultural variation in communication there are fewer examples, but their numbers are increasing. We have already mentioned some examples on assumed cultural differences, for example in gaze behaviour, calls and grooming.

In order to find out whether there is social learning among nonhuman primates there are two ways to go – either to use a longitudinal method, and follow the development of a young infant growing up, or to use a cross-sectional method and compare different groups. Complete longitudinal studies of the development of behaviour in infants are still lacking, but there are many cross-sectional studies on group variation. The most ambitious study of cultural variation is the one by Whiten et al. (1999), who followed the latter method and compared results from different studies on groups of free-ranging chimpanzees. Taken all the years of experiences from fieldwork in chimpanzee sites together, their work represents 151 years of chimpanzee observations. As a basis of comparison, 65 behaviours were listed and the researchers were asked to state for each behaviour if it was customary, habitual, present, absent, absent for ecological reasons, or unknown. Of the 65 behaviours tested, seven were found to be universal (e.g. shake branches) whereas 39 behaviours were assumed to represent cultural variation, since they were absent in some but common in other groups. All groups displayed both unique and shared behaviours. The comparison concerns behaviours in general and it is not specifically focussing on communicative behaviour. Most of behaviours that showed differences across groups in the study were behaviours involving materials, such as use of leaves and sticks to manipulate something in the environment. An example of another character is the hand-clasp grooming which is habitual in Taï, customary in Kibale and in the Mahale groups M and K, but absent in Gombe and Budango. The list of behaviours used in Whiten et al. (1999) was also employed in a group of free-ranging bonobos (Hohmann & Fruth 2003), with the aim of comparing chimpanzees and bonobos. They found that several of the behaviours occurring in chimpanzees, were also common among the bonobos. However, nine behaviours were found only to occur in bonobos, such as groom slap and teeth chatter. The function of cultural differences among nonhuman primates, may be assumed to be the same as among humans, namely to promote group coherence by conformity in behaviours.

5.8 Teaching human language to nonhuman primates

The study of animal communication is a field of study that can be approached from many different aspects, but the dominating perspective is to find out more about the borderline between humans and animals, – what makes humans humans. Traditionally, human language is taken as the main division, and the question about animals learning to use human language has been under hot debate for centuries "as if the only important difference between pigeons and humans, or even monkeys and humans, is language" (Kamil 1998: 20). In this section, we will introduce some of the studies attempting to teach human language to nonhuman primates, by using different means of communication, such as speech, sign language, plastic chips or computer keyboards.

5.8.1 Speech

The idea of teaching spoken language to apes has fascinated people for a long time. We find anecdotal examples of nonhuman primates living in families as pets, and expected to learn to speak, from several centuries ago. The first scientific attempts to teach human language to chimpanzees were home-raising experiments where the chimpanzees were given the possibility to grow up in the same environment as human children. At the age of seven months the chimpanzee Gua was introduced in the Kellogg family and raised together with their own son, Donald, ten months of age (Kellogg 1968, Kellogg & Kellogg 1933/1967). The experiment lasted for nine months. At first, Gua had a faster communicative development but soon Donald surpassed her. To give an example, Gua could comprehend and react appropriately to phrases such as "Where is your nose?" at the age of 14 months, whereas Donald was able to do the same at the age of 16 months. (Interestingly, their pointing behaviour differed, Gua used the index finger, whereas Donald seized the nose between thumb and finger, possibly indicating a different concept of nose, see Chapter 4). Despite a rather good comprehension, Gua did not have any productive vocabulary, and never exceeded her four prime vocalizations; bark, food bark, scream and cry. Another chimpanzee raised in a family was *Viki* (Hayes 1950, Hayes & Hayes 1952). Viki seemed very promising at first. As a baby, she vocalized with babbling grunts and later she was able to produce single words, such as *mama, papa, cup, up*. But despite the fact that she was trained three days a week for six years, she never came to learn more than four words.

Despite the rich environment, with a lot of exposure to spoken language, the experiments with teaching vocal human language to nonhuman primates have not been successful. The reasons that nonhuman primates do not acquire speech are, first and foremost that they lack of vocal control (Fitch 2010), and secondly that the vocal tract configuration in the chimpanzee does not allow the pronunciation of the subtle differences between distinctive speech sounds (see Chapter 3).

5.8.2 Sign language

The introduction of sign language instead of vocal language caused a breakthrough in the experiments of teaching human language to nonhuman primates.

The first chimpanzee to be taught American Sign Language was *Washoe* (Gardner & Gardner 1969). She was raised in the Gardner family and had one member of the research team with her at all times. The teaching method was guidance, molding her hand into the right position, but Washoe also had rich opportunities to learn by observation. After the first year Washoe had a vocabulary of 50 signs and after four years she was able to use 130 signs. She was able to combine the signs in a productive and creative way, similar to semantic relations used by children of Brown's Stage I (Brown 1973: 62, see Chapter 3). Washoe also made the same kinds of simplifications that American deaf children do during an early developmental stage, which indicates similarities in the ability to generalize. However, she never reached the "vocabulary spurt" stage typical of human children around the age of two-three years.

In another project around the same time, Terrace (1979) used sign language to teach human language to the chimpanzee *Nim Chimpsky*. The aim of this project was not only to see how many words a nonhuman primate may learn, but more specifically to investigate if they are able to use syntax. Nim had a staff of over 60 volunteers teaching him, and being around him all waking hours. More than 20 000 multisign utterances produced by Nim were transcribed and analysed. The most frequent patterns are given below.

Box 5.3 The most frequent multiword utterances by Nim

TWO-WORD UTTERANCES	NO OF OCC	THREE-WORD UTTERANCES	NO OF OCC	FOUR-WORD UTTERANCES	NO OF OCC
Play me	375	play me Nim	81	eat drink eat drink	15
Me Nim	328	eat me Nim	48	eat Nim eat Nim	7
Tickle me	316	eat Nim eat	48	banana Nim banana Nim	5
Eat Nim	302	tickle me Nim	44	drink Nim drink Nim	5
More eat	237	grape eat Nim	37	banana eat me Nim	4

These utterances differ from early multi-word utterances by human children in systematic ways. First of all, they do not show the variety of functions found in human children, but are restricted to expressing requests. Secondly, they contain a lot of unnecessary elements; for example in *play me Nim*, *Nim* is redundant. Thirdly, they contain more repetitions, and repetitions of another kind than in human children. What in the surface may look like a four-sign utterance: e.g. *eat Nim eat Nim*, can be analysed as containing two utterances of two signs each: [eat Nim] [eat Nim]. Among the multisign utterances that seem to have syntactic structure involving agent-action-object, we find some with the verb *eat*, *grape eat Nim*, but no more from the list of most frequent utterances. Also quantitatively, Nim's language development differs from human children. Even though his vocabulary showed a steady increase, the utterance length (MLU) did not increase, but stayed at approximately the same level during the data collection. At the age of four, Nim still had a mean length of 1.6 signs per utterance. This is what children use before the age of two years, whereas a four-year-old generally produces utterances of 3.5 or more. Furthermore, Nim's utterances consisted of imitations and responses to teachers. The conclusion is that Nim had no difficulties to acquire new signs, but the ability to combine them was limited – as was his motives to do so.

> I believe that it was Nim's motivation to sign, or his lack of it, that provides the biggest obstacle to the lengthening of his utterances. With few exceptions, it did not appear that Nim had discovered the "power of the word" in the same sense that a child does.
> (Terrace 1979: 222–223)

We will come back to this problem in connection with a discussion of communicative functions and driving forces in Chapter 8. First, two projects of the teaching of sign language to other primates will be presented.

Most experiments with teaching language to nonhuman primates have dealt with chimpanzees. There is, however, also one project on a lowland gorilla (Patterson & Cohn 1990), and another project on an orangutan (Miles 1983). The gorilla, *Koko*, was exposed to two languages simultaneously, American sign language and spoken English, and thus she can be said to have been raised bilingually. The method was to teach sign language explicitly but have spoken English going on naturally. This project started in 1972 and had an unusual design, since it was a longitudinal project, covering a period of ten years. Age of onset was six months for spoken English and one year for sign language. Koko acquired signs in a way similar to children and also showed two vocabulary spurts. Patterson and Cohn (1990:117) concluded that "The data in this article suggest that in many respects gorilla acquisition and production of sign language appears to parallel that of the human child learning language, with the closest parallels to humans learning signs; however, the pace of gorilla language development has been significantly slower than that of the normal human child".

The orangutan *Chantek* (Miles 1983) was 9 months old when he was first taught sign language, in this case, pidgin Sign English, a language with a simplified grammar compared to American Sign English. Only a small staff of caregivers was involved, in order to ensure that they had a personal relationship with Chanek. The language teaching procedure consisted of three stages: first to establish rules for communicative interaction, second to teach gestural signs, and third to use the signs in meaningful communication. Miles reported that Chantak's use of signs was more creative and fluent than that of Nim. Unlike Nim, Chantak did show an increase in utterance length (up to 1.93) and only rarely seemed to imitate the caretaker. The difference is ascribed to differences between species, but also difference in methods. Chantak was engaged in a social relationship with his few caretakers, and therefore more motivated to use signs for communication, whereas Nim had a large staff of people with whom he was unable to build up relationships.

5.8.3 Plastic chips

In another experiment of teaching language to a chimpanzee, Premack (1971) chose to use plastic chips instead of speech or sign language, Premack constructed a system with plastic chips of different colours and shapes, with which the chimpanzee *Sarah* learned to understand and act out rather complex tasks. An example of a sentence is shown in the picture below.

The picture illustrates the sentence "Sarah put the apple in the bucket and the banana on the plate". Each plastic chip has a meaning. There is an arbitrary relationship between the form of the plastic chip and its meaning, except for the chip on top

Sarah. From Premack & Premack (1972)

which means "Sarah". This shows some resemblance of a chimpanzee. From top to bottom the other chips carry the meanings of put – apple – bucket – banana – plate. The intended meaning is that Sarah should put the apple in the bucket and the banana on the plate, which she managed to do.

5.8.4 Computers and lexigrams

Rumbaugh (1977) also used the idea of arbitrary geometric symbols, but implemented them on a computer instead of using plastic chips. The chimpanzee *Lana* was the first to use the artificial language "Yerkish" (named after Robert Yerkes, the founder of the Yerkes Primate Center in Atlanta). It was a system with different symbols, lexigrams. The lexigrams were displayed on the keyboard and Lana soon learned to touch the correct keys to obtain what she wanted (mostly food, but also music). A modified version of the keyboard lexigram was later used to teach the chimpanzees *Sherman* and *Austin* and also the bonobo *Kanzi* (Savage-Rumbaugh & Lewin 1994). The lexigram contained 92 symbols. Sherman and Austin were the first nonhuman primates to make use of intraspecific communication since they were forced to communicate through the computer and also with each other. For example, Sherman could ask Austin to get a certain tool, by pressing the appropriate key on his keyboard.

A breakthrough in the studies of language-learning nonhuman primates came with the bonobo *Kanzi* (Savage-Rumbaugh & Lewin 1994). Kanzi grew up in an environment where it was his mother that was subjected to teaching of symbols, while Kanzi was left to discover it by himself. Kanzi quickly surpassed his mother and became very proficient with the computer. The number of symbols grew to 256, and at the age of seven years, Kanzi was able understand about three-quarters of them, although he only used half of them on a regular basis. Some examples of symbols are *dessert, bowl, monster, coconut, towel, chicken, lettuce, swimming pool, shop* and *observation room* (Savage-Rumbaugh & Lewin 1994:141). At the same time as Kanzi was learning lexigrams he developed an understanding of spoken English. His comprehension skills seem to surpass his production, possibly due to the methodology used. In his own production, however, Kanzi was restricted to the keyboard symbols for expression and had no opportunity to use holistic items, like both first and second language learners do (Wray 2002). There are a number of differences between the situation of Kanzi and the situations of the other nonhuman primates in teaching experiments. First of all, he is not a chimpanzee but a bonobo. Secondly, he starts learning at an early age (like he would in the wild), and thirdly, he is not explicitly taught but learns by observation (also like he would in the wild).

Box 5.4 summarizes the most important teaching experiments. It shows the large diversity in scope and methods.

Box 5.4 Overview of studies teaching human language to nonhuman primates

REFERENCE	SUBJECT	AGE OF ONSET	DURATION	MEDIUM	OUTCOME
Kellogg & Kellogg 1933/1967	Chimpanzee: Gua	7 months	9 months	speech	Some comprehension
Hayes 1950, Hayes & Hayes 1952	Chimpanzee: Viki	newborn	6 yrs	speech	4 words: mama, papa, cup, up
Gardner & Gardner 1969, 1984	Chimpanzee: Washoe	8–14 months	4 years	ASL (Am Sign Lang)	140 signs
Premack 1971	Chimpanzee: Sarah	6 years	4 years	Plastic chips	130 words
Terrace et al. 1979	Chimpanzee: Nim Chimsky	2 weeks	4 years	ASL	Over 125 signs, combinations
Rumbaugh 1977	Chimpanzee: Lana	2 1/2 years	6 months	"Yerkish" computer	Vocabulary to request objects
Savage-Rumbaugh 1986	Chimpanzees Sherman, Austin	Sherman 2 1/2, Austin 1 1/2 years		Lexigrams on computer	Few words, communicative use
Miles 1983	Orangutan: Chantek	9 months	3 years	ASL	56 signs, fluent use
Patterson & Linden 1981	Gorilla: Koko	1 year	10 years	ASL and spoken Eng	More than 400 signs
Savage-Rumbaugh et al. 1986	Bonobo: Kanzi (and other bonobos, chimpanzees)	6 months		Computer and spoken English	Production by using computer Comprehension: spoken English

The projects on language training in nonhuman primates are carried out by different reseachers and they differ in aims, methodology, and outcome. In some cases the main aim is to teach the use of symbols (Viki, Gua, and Washoe), in others the focus is on using syntax (Nim, Sarah) and others emphasize social communication (Chantek, Kanzi). To the differences in use of communicative system (speech, signs, artificial language), we can add differences in learning contexts. Some of the primates in the studies were only expected to reply to requests (Sarah, Lana) whereas others engaged in communicative interaction (Kanzi and Koko). Different methodologies sometimes give different results. Lyn (2012: 360) notes that the study that reported the largest productive vocabulary was the one without explicit training (Kanzi). There is also a difference in age of onset – it ranges between newborn infants and individuals of six years. If there is a critical period for acquisition of a new communicative system in nonhuman primates, this could explain some of the variation in results. Also the assessments differ. For some projects, success is measured in number of words or signs produced, for others success means use of symbols in combinations, or occurrence of social communication – more rarely the use of human language is discussed from the perspective of the natural communication of nonhuman primates.

5.9 Discussion – primary versus secondary communication in nonhuman primates

5.9.1 Primary communication – data base

Like humans may be bilingual and have a first and a second (and a third etc.) language, some nonhuman primates learn a second way to communicate – the human language. In this section we will discuss how they communicate in their primary communicative system and in their secondary system. What do we know about the natural communication in nonhuman primates today? Let us start with the chimpanzees and some quantitative details. Research on captive chimpanzees started in the 1930's, while longitudinal research on wild chimpanzees has been going on since the 1960's, with Goodall's work on the chimpanzees in Gombe and Reynolds' studies from Budongo (e.g. Reynolds 1963, in the very first issue of *Folia Primatologica*) as starting points. Whiten et al. (1999) estimated that the experience of social life in nonhuman primates reached around 150 years of observations. With new sites being added, Whiten et al. (2001) pressed this figure to more than 170 years. McGrew (2004: 92–93) lists 49 sites, in 15 African countries, where wild chimpanzees have been studied. At 17 of these sites, the studies are long-term studies, i.e. contain more than one year of observations. The three subspecies are equally represented: 17 sites contain the Western chimpanzee, 17 the Eastern chimpanzee and 15 the Central chimpanzee. The number of individuals that have been observed is hard to estimate, but from the publications we can infer that it is more than a thousand. At a well-established site like Gombe, the third generation of chimpanzees is now being observed. This seems like a lot of data. But, as is common in empirical research, new data often give new results, and the picture that is emerging is that there is a lot of variation. In a way we seem to know less now than forty years ago – every time a new population is investigated new results have to be interpreted (Strum & Fedigan 2000). Lieberman (1991:155) claims that "We are in the curious situation of knowing more about what chimpanzees can do when they are exposed to human language than about their natural communication".

For other nonhuman primates, the result are scattered. For some species, like the vervet monkeys, we know a lot about the alarm call system, for others, for example the pygmy marmosets we know a lot about infant vocalizations. There is lot of work to be done before we have covered the grounds.

5.9.2   Primary communication – functions and structures

The referential communicative function occurs naturally among nonhuman primates, and is found in alarm calls and food calls. Also the ability to follow gaze indicates that visual communication can be used referentially. Furthermore, studies of audience design provide evidence for an intentional communication.

The four communicative functions suggested by Collias (1960): communication about food, danger, reproduction, and group movements, have been found to occur in the natural communication of nonhuman primates. Most studies on nonhuman primate communication have looked for the referential function in expressions for food and danger, and a few studies have focussed on the social function used in reproduction and group movements. The food issue is expressed acoustically by food calls, and visually by begging gesture. Danger is communicated acoustically by alarm calls. Behaviours having to do with reproduction are usually communicated by tactile and olfactory expressions, and group movements by contact calls. However, studies of teaching human language have not tapped into these different functions or channels, but they have mostly dealt with the teaching of requests, using the acoustic and visual channel. (Requests can be described as having a specific subfunction. They are partly referential, if the request is about a defined object, and partly social, if it is directed towards a specific person).

Nonhuman primates use, just like humans, visual, auditory, tactile, and olfactory channels of communication. We know most about the acoustic expressions. There is a large number of studies where vocalizations have been recorded and subjected to acoustic analysis. These analyses have shown that calls may consist of meaningful parts that can be combined through syntactic rules, just like in human language (e.g. Zuberbühler 2002, Boesch & Boesch-Achermann 2000). There is also a number of studies that have related different calls to different meanings (e.g. Struhsaker 1967, Cheney & Seyfarth 1980) and studies that have demonstrated how vocalizations are learned (e.g. Seyfarth & Cheney 1986, Mitani & Gros-Louis 1998 etc.). To groups with cooperative breeding – with alloparents – a reliable communicative behaviour is imperative, since it involves interaction with many different individuals. These species have more plasticity in their vocalizations than other nonhuman primate groups, and early stages of infant babbling, similar to what is typical to human children, have been identified (Snowdon & Elowson 2001).

For visual expressions, such as body postures, facial expressions, gaze and gestures, there are fewer studies. Some of these behaviours have been discussed in the context of joint attention (gaze, pointing), while others exemplify emotions (body postures, facial expressions). Research on tactile communication has mainly been focussing on grooming behaviour, who grooms whom, and how they do it. The results have shown that in male-dominant species, such as the chimpanzee, it is mostly males that take part in grooming (Goodall etc.). Studies of the structure of grooming have found differences in the hand-clasp grooming (de Waal 2001), and scratching (Nishida et al. 2004). These are assumed to be learned behaviours.

5.9.3 Secondary communication – human language as a second variety

What have we learned by teaching human language to nonhuman primates? One point is that we know that they, like dogs and many other animals, are able to acquire a rather large vocabulary, i.e. labels for events and objects in the external world. This is evidence for referentiality. We also know that they are able to infer meaning in the context. The grammatical abilities are considerably smaller. Their mean length of utterances keep to a level of a two-year-old human child and they do not use own creative structural combinations but repeat what is offered by the interlocutor. But what is language? In the introductory chapter of this book, we used the poem from Saxe to illustrate the complexity involved in defining communication. There has been a bias towards cognitive abilities when chimpanzee linguistic abilities are discussed. Chomsky makes it clear that he thinks of the language training experiments as the teaching of symbolic systems that may tell us something about cognition, rather than communication.

> The study of symbolic systems taught to apes will no doubt prove rewarding in mapping their intellectual capacities, and perhaps it will ultimately help to locate more exactly the cognitive systems that are specific to humans, a contribution of some potential interest.
> (Chomsky 1980: 440)

Although Sue Savage-Rumbaugh has a totally different view on linguistic capabilities in human and nonhuman primates, she shares the view on language as a feature of cognition (Savage-Rumbaugh et al. 1985).

> The question has fascinated a number of psychologists ever since the discovery of apes by western civilization. The behaviour of these animals seemed so intelligent that many scientists were repeatedly puzzled as to why they could not learn to speak.
> (Savage-Rumbaugh et al. 1985: 177)

5.10 Summary

In summary, the findings about natural communication in nonhuman primates demonstrate that both referential and social functions are used. Most research has focused on acoustic communication, and report findings about calls associated with referential functions, such as alarm calls and food calls. Nonhuman primates live a fission-fusion social life, and use a lot of social communicative functions as well, greeting and pacifying communicative expressions; acoustic, visual, tactile and chemical. They groom each other and synchonize vocalizations and gestures. Species with cooperative breeding use more vocalizations than others, and these vocalizations often show gradual development. Phonological analyses show that there are discrete elements in the vocalizations, and some nonhuman primates have categorical perception comparable to human newborns. The different elements can be combined into patterns and it has been suggested by some that nonhuman primates may have rudimentary syntax.

When it comes to teaching language to nonhuman primates, it is the referential function that has been in focus. Like many other animals, nonhuman primates seem to have no difficulty in acquiring and producing words by sign language and keyboard strokes, to refer to some external phenomenon. Efforts of grammar teaching have been less successful, but the results have shown that nonhuman primates are able to produce combinations of words and to comprehend certain sequences of signs. However, nonhuman primates that are taught human language use it mostly to request objects, and not to cooperate and do things together. Futhermore, they rarely transfer the knowledge to others. This may imply that they lack contexts where referential language is needed – they already have expressions for social purposes.

5.11 Suggested readings

Cheney, D. L. & Seyfarth, R. M. (1990). *How monkeys see the world.* Chicago: University of Chicago Press.
   In this book, the authors ask a unusual question: "What it is like to be a monkey?" The social behaviour of monkeys living under natural conditions is described with the aim to understand the underlying motives and strategies. The book is mainly based on data from the authors' own research on vervet monkeys but also studies from other nonhuman primates are discussed.

Hallix, W. A. & Rumbaugh, D. M. (2004). *Animal bodies, human minds: ape, dolphin, and parrot language skills.* New York: Kluwer Academic.
   This book builds on a cooperation between an psychologist and a biologist. It describes the projects on teaching language to animals in great detail, and discusses how far we can expect to go in this endeavour in future studies.

Pika, S. & Liebal, K. (2012). *Developments in Primate Gesture Research.* Amsterdam: John Benjamins.
   This is an edited book that gives the state of the art in nonhuman primate gesture research. The chapters describe gestures in different species such as chimpanzees, gorillas and mandrillls. Many chapters focus on ontogenetic development of gestural behaviour. Particularly interesting is the use of conversation analysis designed for human speech to analyse requesting behaviours in chimpanzees.

Chapter 6

Man's best friend – the dog

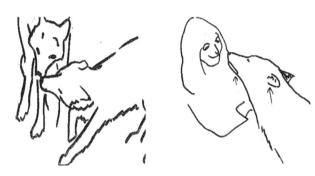

6.1 Introduction

As mentioned in Chapter 1, comparisons between humans and nonhumans are often motivated by a genetic relationship between the species (homologies), but they can also be motivated by a similarity in behaviour (analogies). The reason for including a chapter on communication in dogs in this volume is the second one – the similarities. It is not similarities in *body structure*, which we share with the other primates, but similarities in the *social structure*. The ancestor of dogs as well as the ancestor of modern man, are both assumed to have lived in tight groups, with coordinated hunting and cooperative breeding. These initial similarities in social behaviour have become more and more pronounced during domestication. The pictures above illustrate how a wolf greets a conspecific, and how a dog greets a human. In both cases the animal pays respect by drawing back the ears, half-closing the eyes and attempting to lick the corners of the mouth of the other participant.

Dogs have lived with humans for at least 15 000 years (Savolainen 2007). During this long period of shared lives, dogs have adapted their behaviour to human social life and the inter-specific communication between humans and dogs is very successful. Because of this, dogs have sometimes been regarded as an artificial species, not suitable for ethological research. However, for the purpose of his book, the fact that dogs live so closely together with humans makes the study of their communicative behaviour even more interesting, since studying dog-human communication may be a way to understand our own communication and how it evolved (Hare & Tomasello 2005, Jensen 2007).

In this chapter we will present research on dog communication both in intra-specific contexts (between dogs) and inter-specific contexts (between dogs and humans). In the discussion of intra-specific communication we will also include research that has been conducted on wolves. First some words about the species and why it is important to bring in wolves when talking about dog communication.

6.1.1 The Canine class

The domestic dog (*Canis familiaris*) belongs to the *Canidae* family. This family includes around 35–40 species, for example wolves (*Canis lupus*), coyotes (*Canis latrans*), jackals (*Canis aurerus*) and foxes (*Canis rufus*). Some of these live in social groups (wolves, coyotes and jackals) whereas others are solitary hunters (foxes). The ancestry of dogs has been discussed and different suggestions have been put forward. In the 1950's Konrad Lorenz suggested that some dogs came from wolves and others from jackals (Lorenz 1954). Today however, most researchers agree that all dogs come from a wolf species, possibly extinct. Evidence from fossils date the first domestic dog from around 15 000 years ago in East Asia, from where the dog spread all over the world (Savolainen 2007).

The domestic dog displays an exceptional variation in size and other characteristics. No other land mammal has a similar diversity (Spady & Ostrander 2008). Their sizes range from the tiny Chihuahua (height 6–9 inches/ 15–23 cm, weight 2–6 pounds/ 1–3 kilos) to the large Irish wolfhound (height 30–36 inches/ 85–90 cm, weight 105–135 pounds/ 47–71 kilos). Other features that differ and are assumed to have genetic basis are the snout, the ears, the tail and the coat. The snout can be long and pointed, or short and flattened; the ear can be hanging or standing; the tail can be long, short, curled or double curled.

The International federation of Kennel Clubs (Fédération Cynologique Internationale FCI) recognizes over 300 different breeds. The number is not fixed, but new breeds are regularly established. The breeds are categorized in groups according to their functions. For example, the British Kennel Club differentiates seven groups: hounds, gundogs, terriers, utility dogs, working dogs, pastoral dogs, and toy dogs.

The hound group consists of dogs that are used for hunting a quarry by following it. They do so either by scent or by sight. Scent hounds, such as beagle, dachshund and bloodhound, hunt by following a trail while indicating the direction of the prey by barking, whereas sight hounds, like greyhound, borzoi and whippet run after the target and kill it. The physiology of scent and sight hounds differ a lot. The beagle is sturdy and strong. The name "beagle" derives from French *beegueule*, meaning "screamer", indicating their loud barking. The nose is large, the lips are moist and loose and the ears are long – these are features that makes it easier to capture and protect the scent. They have their most extreme expression in the bloodhound, who has big nostrils, very long thin ears and skin in the face and around the lips so loose that it hangs in wrinkles. In contrast to the scent dogs, the sight dogs have tight lips, long necks, long legs and deep chests. The greyhound runs with a 'double flight' and is one of the fastest dogs in the world (Coppinger & Coppinger 2001). The greyhound has a sharp eyesight and is able to detect movements easily, which makes it possible for it to hunt in the open by running after the prey, for example deer and hare, capturing and killing it.

Gundogs help the hunter to find live pheasants, quails and other birds by pointing at the quarry (pointers, setters) by springing at it (spaniels), or by retrieving game that have been shot (retrievers). Pointers got their name from the pointing behaviour. While searching, these dogs run in zig-zag with head up against the wind in order to find the scent of the prey. When they find the prey they freeze into a pointing position, with muzzle in the direction of the prey – they "point" at the prey. Originally hunters threw a net over the prey that the pointer had found, but today guns are used. With its large nostrils and concave muzzle the pointer is the perfect model to catch the air scent. The behaviour of retrievers are also in their name: they fetch and bring the bird to the human hunter. Retrievers have many features that are suited for work in water, for example the oily, water-repellent coat.

The name "*Terrier*" comes from Latin *terra*, meaning earth, and it refers to the ability to dig into underground passages and drive out vermin, such as rats, mice and foxes. Most short-legged terriers come from Scotland (e.g. Scottish terrier, cairn terrier, West Highland white terrier) where they have become specialists in digging into burrows and cairns – hence the name Cairn terrier. There are also long-legged terriers, for example fox terrier, Irish terrier, Welsh terrier, and Lakelandterrier (from the Lake District). Terriers are known to be extremely brave and fierce.

The group with *utility dogs* is a mixed group with dogs that do not have specializing. Some examples are chow chow, dalmatian, keeshound.

Working dogs are dogs that work with pulling sleds, guarding and rescuing, for example Siberian husky and St Bernard. Sled dogs are known to run fast and to be able to cooperate in a team. They have a double coat and can cope with cold of minus 75 degrees. Their bodily structure is more expressive than many other dog breeds, with upright ears, long tail and furred facial markings. This facilitates communication and diminishes the risk of unnecessary fights in the team. The Saint Bernard got its name from the monk Bernhard de Menthone, who founded a hospice in the Aosta Valley in the Italian Alps in the 11th century. The Saint Bernard breed was developed from various large dog types in the passes between Italy and Switzerland. The most famous representative is Barry who is said to have saved many lives and rescued many lost wanderers. Barry died in 1814 and has a monument in Cimetiére de Chiens in Paris dedicated to him. The Saint Bernhard is one of the largest dogs in the world.

Pastoral dogs are dogs that working with livestock. There are two different types with different manners of working: one type of dogs herds the animals, and the other protects the animals from thieves and predators. In the first group we find for example the border collie, Australian kelpi, Welsh corgi, Swedish vallhund, all energetic dogs with swift movements. The herding dog cooperates with the shepherd to move the flocks and to bring them in to the owner. Some dogs work from the front and others from behind the flock. The border collie typically gets in front of the sheep, crouches down and "eyes" them by staring intensely (it is noteworthy that some border collies have blue eyes, and some one brown and one blue eye, which may help in giving he stare a hypnotizing character). The Australian cattle dog works from the back: they have the same manner as the Australian dingo to sneak up behind the animals and bite their heels, and they are therefore called "heelers" or "biters".

A different kind of pastoral dog is the guarding dog. These dogs are generally large and strong, usually white in colour, and capable to attack big predators such as wolves and bears. Examples are kuvasz, komondor and Maremma sheepdog. They are often left in charge of the flock, and they have a strong guarding instinct. This trait, together with a willingness to stay with the flock, makes them suitable also to watch over endangered species. For example, in Warrnambool in Southern Victoria, Australia, a project to guard penguins from predators by using the Maremma sheepdog as protection has proved successful, and more projects are underway (http://www.warrnambool.vic.gov.au/index.php?q=node/943; retrieved Nov 24th 2012).

Finally, the *toy group* includes small companion dogs, originally bred for various purposes (sometimes sacred) but functioning as family pets because of their small size, for example Chihuahua, Pekingese. The Chihuahua dog is the world's smallest dog. It originates from Mexico, where it was used in religious ceremonies among the Aztecs who regarded it as sacred. Also the Pekingese, or the Lion Dog, has a background as a worshipped semi-divine. It was considered to look like a lion and only royalty was allowed to own it. Its appearance represents the total antithesis of the wolf, and none of the features that are used for wolf communication, for example, the upright ears, the movable tail, the nose or different gaits are present.

The different dog breeds meet the needs of humans in varying contexts. However, even if some breeds seem to be predetermined to perform certain tasks, the behaviour does not follow automatically. Not all pointers are good at finding prey, not all beagles bark when they follow a trail, not all collies are interested in collecting the sheep, etc.

When different dog breeds exhibit different behaviours it means that the animal has the structure and disposition for certain behaviours – but there is also room place for learning and adjusting. The behaviour itself is not passed on from generation to generation, but the potential is there "and it is the handler's job to release the potential" (Coppinger & Schneider 1995: 26).

Basically, all behaviours displayed in different dog groups can be traced back to the grey wolf. The different phases in wolf hunting are used for partly other purposes in dogs; how hounds follow the prey, collies herd the sheep, gundogs fixate, flush and retrieve, terriers grab and bite, and working dogs pull and guard. By human interference, dogs have become limited specialists of their respective areas instead of the all-round hunter that is the hallmark quality of the wolf.

6.1.2 Social life of wolves and dogs

The intraspecific social life in dogs is difficult to investigate, since most dogs spend their lives close to humans, and not among conspecifics. There is a substantial body of research on the social life in wolves, and these studies are commonly used as baseline data to understand the dog. The first wolf reports appeared around the 1940s (Murie 1944) and then many other studies followed, mostly by researchers who observed natural behaviour in the wild (e.g Crisler 1959, Mech 1966, Mowat 1965).

Wolves live in extended family units, so-called packs. The pack consists of a breeding couple, also known as the alpha couple, their offspring, some juveniles from an earlier litter, and often some adult subordinate wolfs. Pack sizes vary between five and fifteen members. A large part of the life is dedicated to hunting – searching prey, chasing and killing. Unlike the domestic dog, where different breeds are specialised in different types of prey, wolves are generalists and hunt all kinds of prey, birds and rabbits as well as elk, reindeer and bison. Their hunting behaviour is highly adaptable and they hunt individually as well as in group. When the prey consists of large animals, the wolves coordinate their efforts and use the strategy to separate the old and fragile members from the others. After the killing, the meat is divided between the members. Before they are old enough to participate in the hunt, juveniles stay at a rendezvous place where members of the pack feed them after the hunt.

Communication is of vital importance in the wolf pack. There are many expressions for submission and dominance and much time is spent on social play, with mock attacks and reconciliations. Usually it is only the alpha couple that mates and reproductive activity is not socially accepted among the others. The flexibility in wolf social behaviour, however, is shown by the fact that when there is plenty of food sometimes more than one litter of offspring may be born during the year. Another flexible feature is that maturation among the young females which is between 2–3 years of age in the wild, can occur at the age of only one year in captivity with good nutrition (Medjo & Mech 1976).

The upbringing of young cubs is a good example of cooperative breeding (see Chapter 1.3). In cooperative breeding other individuals than the mother (alloparents) assist in caring for the infant. During the first weeks, the mother stays in the den nursing the pups, while the other wolves do the hunting and bring food to her. The young pups are in a vegetative phase during these first two-three weeks, unable to see or hear anything. They stay silently close together when left in the den for short periods. When the pups are weaned, it is the responsibility of all pack members to supply half-digested food to them by regurgitation, and also to baby-sit when the mother is away hunting. This means that the pups get used to having different caretakers and communicating with different individuals. Around the age of two months, the pups move from the den (which may be used for generations of wolf litters) to a so-called rendezvous site, where they stay and play while the adults are hunting. The first year of life is spent in this way, getting help from different pack members, and observing and learning from them (Busch 1995).

Other social Canides, for example coyotes and jackals also live in extended family groups, and share responsibility of feeding the young. The life of domestic dogs, however, is markedly different, and it varies according to how much they depend on humans. Some dogs live with only limited contact with humans. These dogs are usually characterized as stray dogs and they lead a free-ranging life, joining others temporarily when there is an oestrus female or a shared food source.

There are both similarities and differences between the social life of wolves and the social life of the free-ranging dogs. In a study of free-ranging dogs in the Indian town of Katwa, Pal (2005, 2008) reports that the mothers stayed with the pups for a total of 10–13 weeks, first nursing them and then feeding them by regurgitation. Half of the fathers that were observed (4 out of 8) helped in bringing up the pups by guarding them and feeding them by regurgitation, and sometimes two neigbouring mothers nursed each others' pups. This means that these domesticated but free-ranging dogs had similar alloparenting behaviour as wolves.

Another kind of free-ranging dogs are the so-called feral dogs. These dogs have no contact at all with humans. They live in social groups similar to wolf packs, with some significant differences: feral dogs do not have the social mechanism that controls reproduction but litters can be produced many times a year, also at times when it is difficult to find food. Another problem for feral dogs is that the female often goes into isolation when it is time to give birth and takes care of the offspring herself. With no cooperation in taking care of the young, the result is a high mortality among puppies (Boitani, Ciucci & Ortolani 2007, Miklósi 2007a, see also Zimen 1971, 1981).

Domestic dogs live a social life that bears some similarities to the life of wolves and other Canids, but which differs in being inter-specific rather than intra-specific. Many dogs prefer human company to company with other dogs (Feddersen-Petersen 2007), and many dogs do not even have a choice, living in a human family that can be said to have replaced the Canid family. Humans feed the dogs, humans decide when it is time for mating and whom to mate with, and humans cooperate with the dog in handling the offspring. Unlike young wolves, young domestic dogs socialize with two species. Firstly they identify with their own species, and secondly they socialize with humans, who can be seen as taking on the roles of the pack members in the wolf pack. Once dogs have learned to treat humans as belonging to the group, they get more and more attuned to their human family and learn to interact and comprehend human communicative behaviours. The adaptation is reciprocal; many owners treat their dogs as family members and even address them in the same way as they talk to young children (cf. 6.8).

6.1.3 Observation techniques

Studies of wolves are conducted either in the wild or with captive individuals. As mentioned above, pioneering work in the 1940s gave the first insights in the social life of wild wolves. However, detailed studies of the repertoire of communicative behaviours in the wolf pack are scarce. This is due to the fact that wolves travel long distances and it is difficult to set up cameras close enough to capture the dynamics of wolf-to-wolf interactions. Instead, the study of wolf communication is usually done on captive individuals and often focussing on comparisons to the behaviour of dogs under similar experimental conditions (e.g. Pulliainen 1967, Scott & Fuller 1965, Zimen 1971).

When it comes to studies of domestic dogs there is a problem in deciding what is the natural environment. There are some studies on feral dogs, but it is uncertain if feral dogs should be regarded as the best representatives of dogs. Instead, the human family may be the most natural social unit for the dog. Many studies of dog communication have focussed on inter-specific communication; namely communication between humans and dogs. Often the studies are based on experiments where the dog is expected to react to a certain human behaviour, e.g. pointing. These experiments are video recorded and analysed by frame-by-frame analysis.

Vocal communication in wolves and dogs has been recorded and analyzed using different techniques. It is particularly the chorus howling in wolves that has attracted attention. Howls have been carefully described with respect to frequency ranges, harmonies and length. More recently, acoustic analyses of dog vocalisations have also been undertaken.

As mentioned before, the development of the playback technique was an important step towards controlled studies on animal vocalizations and their associated meanings. To record vocalizations and play them out to individuals of the same species, makes it possible to interpret potential meanings. However, despite the successful use of playback in many species (e.g. insects, elephants, vervet monkeys), the method has not been applied much in studies of wolves and dogs. One area where it has come to use is the investigation of chorus howling in wolves. Playback of howling often results in howling responses by other wolves, which suggests that it is contagious, but also context-dependent, since only wolves that have a territory to defend howl back, whereas other wolves do not answer (Harrington & Mech 1978). Other examples of playback are studies of how dogs react to growls of varying frequences (Taylor et al. 2010) and of how humans interpret dog vocalizations (Pongrácz et al. 2005).

6.2 Functions of Canine communication

6.2.1 Social functions

The social structure of the wolf pack with dependence on group members puts high demands on successful communication. The wolves constantly emit signals about their emotional and physiological states, signals about dominance and submission. The same expressions that are seen among wolves can also be found among dogs. In the event of a conflict, the parties would be eager to make up. Studies have shown that reconciliation after conflicts is an important part of social behaviour both in wolves (Cordoni & Palagi 2008) and in dogs (Cools et al. 2008). This is a fundamental characteristic in groups where individuals need to cooperate.

As was indicated in previous sections, dog breeders have focused on the use of dogs for human benefits. Communicative behaviour between dogs has not been of interest and it has rarely been selected for. A context where intra-specific communication is important is in the teamwork of dogs that pull sleds, such as the Siberian husky. Their bodily structure resembles that of wolves – upright ears, long tail and furred facial markings. This makes communication easier and diminishes the risk of fights in the team.

Despite differences in social structures, the social functions of communication have the same expressions in wolf packs and dogs living in human families. This makes is easy to study the social functions of dogs' communication (and we know more about the social communication of dogs than of social communication in nonhuman primates). Dogs show non-aggression by wagging the tail, trying to grip under the cheek and lick the corners of the mouth of the leader. This is the same irrespective of the leader is another dog or a human. After conflicts, many dogs want reconciliation – this is sometimes interpreted by humans as if the dog is ashamed of what it has done. An important behaviour that helps prevent aggression and jeopardize cooperation is the social play, and this is common between dogs as well as between dogs and humans. If a dog would try to take over leadership and dominate one or many family members, this is shown by expressions such as raised fur and growling, which is easily understood by the other members.

6.2.2 Studies of the referential function

Studies of the referential fuction in wolves and dogs mostly come from human-dog interaction. However, there seems to be a potential for referential communication in the intraspecific behaviour of wolves. As is pointed out by Miklosi and Soprani (2006) referential communication is often associated with cooperation; directing another's attention to something. In a species with coordinated hunting some referential communication could therefore be expected. The wolves' hunt has been described as having five phases: locating prey, stalk, encounter, rush and chase (Peterson & Ciussi 2003). The stalking phase contains a moment of restrained waiting: "Although they seem anxious to leap forward at full speed, they continue to hold themselves in check" (Mech 1970:199).

A phase of "restrained waiting" is common in dogs hunting with humans. Pointers (and other gun dogs) are known to keep a "freezing" position when they have targeted the prey. This position is interpreted as a "point" by humans, and also by dogs nearby. In fact, this behaviour is specified in the manuals of field trials for pointers. A high-quality gun dog will acknowledge the point of another dog by stopping and "sharing the point". In bushy terrain it is a further merit if the dog leaves the prey and goes to fetch the owner to "report" before returning to the place where the game is squatting (Christoffersson & Bärg 1990). This behaviour is also used in the family, when the dog wants to draw the owner's attention for example to an empty water bucket (cf. Miklósi et al. 2000). It has been suggested that breeds that are used for certain hunting and herding functions are particularly good at communicating with humans (McKinley & Sambrook 2000, Wobber et al. 2009). If that is true, it is not surprising that gun dogs perform referential functions in their own communicative signals.

Most studies on referential function in wolves and dogs are focussing on how they comprehend human pointing. The results show that dogs (already at the age of 6 weeks) understand a human hand pointing without explicit training. In comparison to nonhuman primates dogs have been found to be more skilful than the primates (Hare et al. 2002). Dogs also outperform wolves at understanding human pointing

(Miklósi & Soprani 2006). Even if the wolves get training, they will not perform as well as dogs. When testing for competition of visual and olfactory signals it has been found that dogs rely more on the human pointing to a bowl than the olfactory cue telling that the food is placed in another bowl. Only if the dog actually sees the food it will become uncertain about the human pointing (Szetei et al. 2003).

Another area where the referential function is tested is in studies of word learning. Dogs are able to learn labels for objects in a natural situation and can show that they understand the words by fetching objects and bringing them to different persons (Kaminski et al. 2004), and to touch a keyboard with arbitrary signs to communicate requests (Rossi & Ades 2008). Guide dogs with blind owners adjust their communication to the blind context, by using licking sounds to get attention. This shows that dogs are able to accommodate to varying conditions, a so-called audience effect (Gaunet 2008). In a study where the dogs were asked to fetch an object, and the object was indicated either by pointing or word label, there was a tendency for the dogs to find the word label more reliable than the pointing (Grassman et al. 2012). The box below summarizes some selected studies in the area of referential communication.

The studies summarized in Box 6.1 illustrate the wide span of research questions and methods used in the purpose of finding out about the dog's ability for referential communiction. Data from 260 dogs of different breeds and ages are used. The research questions range from the importance of synchronizing gaze and hand, different hand movements, distractions of odour and commands, learning of words and lexigrams, to the dog's behaviour in contacting humans. The results show that dogs are able to follow both hand and gaze pointing, they rely on pointing even when there is conflicting information of odour, they have a good capacity to learn words and lexigrams, and they are able to find ways to communicate with humans by gaze or other means.

6.3 Acoustic communication

Acoustic communication may reach over long distances and it is useful also in dark places such as the pen where the mother stays with the newborn pups. Dogs and wolves have different vocalizations for different purposes. Some are close-range: barks, growls, whines and yelps; others are long-range: howls. Box 6.2 illustrates some of the vocal expressions and their functions/contexts where they occur.

The short-range vocalizations are used mostly to maintain the social relationships within the pack, whereas the long-range vocalization, the howl, is used to advertise presence to outsiders. Dogs exhibit many of the same basic acoustic communicative patterns as wolves, with some exceptions: barking is more important, and occurring

Box 6.1 Examples of studies investigating Canine referential communication

TYPE	SUBJECTS	METHOD	MAIN FINDING	REF
Human pointing	14 dogs	Experimenter points at food while looking at the food or at the ceiling	When compared to results from children and chimpanzees, dogs are more similar to children than the chimpanzees are. They differentiate between look at the food and look at the ceiling – chimpanzees don't	1
Human pointing	10 dogs	Different arm gestures, e.g. reverse pointing	Evidence for referential pointing: Dogs understand pointing and can generalize it to new pointing gestures, such as pointing with a stick and pointing across the torso	2
Human pointing	9 wolves, 17 dogs	Distal pointing (50 cm) at food	Dogs detect referential functions at 4 months without training but wolves need extensive training	3
Human pointing	55 dogs, 11 chimps, 12 wolves	Different arm gestures	Dogs are better than chimpanzees at understanding pointing, dogs are also better than wolves, dog puppies understand pointing already at age 9–12 weeks	4
Human pointing	64 dogs	Different arm gestures	Dogs follow human pointing already at 6 weeks	5
Human pointing	55 dogs	Pointing vs. odour, vs. seeing the food	Dogs follow human pointing even if there are conflicting cues by odour. Conflicting visual cues more disturbing	6
Human gaze	23 adult dogs	Commands with owner facing dog or looking away	Dogs are sensitive to human head orientation and gaze	7
Human spoken words	The dog "Rico"	The dog was taught 10 new labels for items	Rico could correctly identify and retrieve the new items after a single exposure. Performance comparable to 3-year old child and evidence for fast mapping	8
Combination of words and pointing	"Paddy" "Betsy"	Both naming and pointing were used in a fetching task	Word-trained dogs rely more on the naming of the object than on the human pointing	9
Lexigrams	"Sofia"	Touch keyboard lexi-grams with paw	Dogs are able to learn arbitrary visual signs and use them to solicit things (e.g. water, toy, walk)	10
Dog showing	10 dogs	Experimenter hides food, owner enters	Dogs communicate location of food to owner with gaze alternation	11
Dog audience effect	9 guide dogs, 9 pet dogs	Dogs asked for help to retrieve food	A new vocal behaviour in guide dogs. They use sonorous mouth licks in order to attract the attention of their blind owners	12

References: 1. Soproni et al. 2001   2. Soproni et al. 2002   3. Virányi et al. 2008   4. Hare et al. 2002   5. Riedel et al. 2008   6. Szetei et al. 2003   7. Virányi et al. 2004   8. Kaminski et al. 2004   9. Grassman et al. 2012   10. Rossi & Ades 2008   11. Miklósi et al. 2000   12. Gaunet 2008.

Box 6.2 Vocalizations (after Feddersen-Petersen 2007, Serpell 1995, Yeon 2007)

VOCALIZATION	CONTEXT
Bark	Alert/warning
	Territorial defense/rivalry/defense
	Individual identification
	Play solicitation
	Greeting, call for attention
Growl	Offensive and defensive warning
	Play
	Defence
	Warning/threat
Whine	Greeting
	Frustration
	Active submission
	Attention seeking
Yelp	Pain, great stress
Howl	Territorial maintenance
	Individual recognition
	Hunt solicitation
	Reactive (in response to sirens etc)

in a wider range of contexts in dogs, whereas the long-range vocalization, howling, is less prominent (Yin & McCowan 2004). In accordance with the "frequency code" (Morton 1977), the high-pitched vocalizations (whines, yelps) express non-aggression and low-pitched vocalizations (growls) express aggression.

6.3.1 Barking

Barking is the most prominent vocal expression among dogs – and it differs in many ways from the kind of barking in wolves. Miklósi (2007b: 219) writes that "dogs 'invented' the harmonic version of the bark, not recorded in the case of wolves. This means that barks are used to express a much wider range of emotions in dogs in comparison to wolves". The dogs bark for a range of different purposes, for example alert, defense, individual identification, social play, and contact request. There is not much experimental research on dog barkings, due to "the huge variability in size, anatomy and behaviour in the species. It is impossible to investigate in an ethological experiment any sample of dogs that would represent the whole spectrum of dog-morphs"

(Pongrácz et al. 2010:142). The size of different dog breeds varies enormously, from the chihuahua to Saint Bernard, and the correlation between body mass and vocal tract length results in differences in the acoustic signal (Riede & Fitch 1999).

Structurally, most dog barks can be placed at a continuum between harmonic and noisy sounds, with many mixed sounds. Disturbance barks (warning, defense) are harsh, low-pitched and relatively long, whereas play barks are more tonal and have a higher frequency. Both these types are produced as a sequence, while the islolation barks ('I am lonely') typically are produced as single sounds (Yin & McCowan 2004). In comparison to dogs, wolves have a more restricted repertoire of barks.

Barking has sometimes been assumed to be a neotenic feature, i.e. that the domesticated adult animal should exhibit juvenile characteristics. This is something that often happens in domestication. An interesting hypothesis is that the barking among dogs also represents a social learning in the context of human society. Many animals modify their communicative behaviour to the social group, and since dogs have stayed close to humans for a long time they may be adapting to a human behaviour. The fact that dog barks are similar to human speech in terms of syllable length suggests that barking could be seen as an adaptation to living with humans (Fukuzawa et al. 2005, Feddersen-Petersen 2007).

6.3.2 Growling

Growling is the most aggressive expression in dogs (as well as in wolves) and it is produced at extremely low frequencies. As mentioned earlier, body size is an important variable in social interactions and the correlation between body size and acoustic signal is usually straightforward: large mechanisms produce low frequencies. By vertically raising the lips, wrinkling the nose ridge, the dog and wolf enlarge the vocal cavity and produce sounds with low frequencies – a growl. The opposite function is reached when the lips are drawn back horizontally since this decreases the resonance in the vocal cavity and the sound indicates a smaller body size. Dogs are able to perceive and react to this difference in size when they are exposed to playbacks with growls of different frequencies (Taylor et al. 2010).

6.3.3 Chorus howling in wolves

The most conspicuous acoustic signal among wolves is the *chorus howling* (Harrington & Mech 1978, Mech & Boitani 2003, Zimen 1981). Similarly to many other animals with acoustic signals, such as birds and frogs, the wolves howl mostly in the morning and evening. This is a time of the day when the meteorological conditions are most favourable for sound transmission. The howls usually start with a few barks from one or two wolves and then the whole pack joins in, the adults having deeper voices and the younger howling in high-pitch tones. The main functions of the howl are assumed to be the maintenance of territory border, individual recogintion, socilitation to hunt and reaction to a sound (Harrington & Mech 1983, Mech & Boitani 2003). Acoustic analyses of howls show individual differences in pitch range and pitch change in the howl. Experiments with playback to wolves have shown that wolves react differently to different howls, which indicates that they are able to recognize the voices of the individual pack members.

Although there are individual features in the howls, no geographical variation has been observed between wolves in different areas. This suggests that unlike many other animals, wolves do not develop dialects in their vocalizations. A study comparing howling in European and North American wolves revealed striking similarities, despite the fact that these two groups have been separated since the closing of the Bering land bridge 10 000 years ago. Due to genetic differences the Iberian wolves have been suggested to represent a subspecies, *Canis lupis signatus* (Wayne & Vilà 2003). However, these genetic differences do not seem to be paralleled by differences in howling structures. The analyses from howls in the Iberian Peninsula showed that they are very similar to howls from North American wolves, except for some variation in the breaks (Palacios et al. 2007).

6.4 Visual communication

Visual communication may differ between dog breeds because of their physical appearance. The main idea is however to signal dominance with an upright body posture, high head, and high tail, and to signal submission by doing the opposite. It is interesting to note that it is not the real difference in size that matters: a small dog can dominate over a larger dog by behaving fiercely.

Not only postures but also movements are expressive. In most dog breeds the gait and the tail wagging reveal the inner mood. A particularly expressive movement is the play bow, which is a strong signal inviting play behaviour.

6.4.1 Head and face

Eye contact is important in communication and also the facial expression plays a vital role. The wrinkled nose and bare teeth are typical expressions both in aggressive and in submissive situations, but in aggression the nose is more wrinkled and only the front teeth are visible, whereas in submission all teeth can be seen. These expressions are associated with vocal expressions. In aggressive situations a low frequency growl is produced with the corners of the mouth less retracted that in submission context, shaping the larger cavity mentioned above. To produce the submissive high frequency sound, there is a smaller vocal cavity and also fewer nose wrinkles.

Wolves have conspicuous facial expressions, which are accentuated by the colours on the coat in the face. However, during domestication other characteristics have emerged in the dog. These features have increased the differences between dog and wolf, both in appearances and in communicative behaviour. The expressions used by wolves are not possible in all dog breeds. For example, dogs with long fur, like the komodor, or dogs with wrinkled faces, like the boxer, will have to use other means of communication than facial expressions. In a study comparing behaviours in different dog breeds, the number of communicative behaviours was found to correlate with degree of physical similarity with the wolf. Fifteen different behaviours used by wolves in aggressive situations were chosen and investigated in ten different dog breeds. The breed whose exterior was considered being the most dissimilar to the wolf was the Cavalier King Charles spaniel and, as expected, individuals of this breed used only two of the selected behaviours. The Siberian husky, on the other hand, which was the breed most similar to the wolf, exhibited all of the wolf behaviours (Goodwin et al. 1997).

6.4.2 Tail

In his pioneering work on wolf communication, Schenkel (1947) described the tail as being the most forceful element of visual communication. He suggested a model along a continuum that range from self-assertion with the tail high and stiff, through different degrees of tail lowering and wagging, to active submission with the tail tucked under the body.

Dogs that have a moveable tail use it to signal dominance and compliance. Both the posture and the movements are important. The importance of the tail was shown in an experiment on robots by Leaver and Reimchen (2008). They constructed a robot dog of black synthetic fur, in a standing position and with a tail which could be controlled. The dog had a shoulder height of 50 and the appearance of a Labrador retriever. Two types of artificial tails were used, one short (9 cm) and one long (30cm), both in a vertical position. The robot dog was placed in areas where dogs are let free to play, and a camera was positioned 10 m away from the model. Almost 500 dogs encounters were recorded. The behaviour of the approaching dog was scored according to: speed of approach, position of tip of tail, position of base of tail, head position and tail motion. The fact that 72% of the dogs first went to the tail region, was taken as evidence for the robot dog being a realistic model, since this is what happens in real dog-dog encounters. The results showed that the dogs of the same size or larger were the ones that most often approached the robot. Smaller dogs often continued past without approaching, or stopped at some point. Most approaches were observed when the robot dog had the long wagging tail, which indicates that the strongest social cue for approaching strangers is a long wagging tail. Anecdotal evidence suggests that dogs with cut tails have more difficulties in their social communication, and they (for example the boxer) wag the whole rear end when showing friendliness.

6.5 Tactile communication

Tactile communication is present already in newborn puppies. They prefer huddling together with close body contact. This close contact reduces stress in both dogs and humans, measured by heart rate and blood pressure (Vormbrock & Grossberg 1988). Other means of tactile communication is touching with the muzzle, licking and pressing the whole body against somebody. Expressions for submission and dominance abound. The gentle seizing of the opponent's muzzle means submission if it comes from below and dominance if it is from above. Zimen (1981) registered around six muzzle-to-muzzle or muzzle-to-coat contacts per hour per wolf, and concluded that the "constant reciprocal contact furthers the cohesion of the pack" (Zimen 1981:76).

Mock threats and submission signals are used extensively in play situations – both among wolves and among dogs. When young puppies play with their parents the adult would seize the puppy's muzzle from below.

6.6 Chemical communication

Olfaction, the sense of smell, is probably much more important in the world of Canines, than in humans. Wolves and dogs have approximately 150–200 million olfactory cells, which can be compared to the 5 million in humans.

Markings of urine and faeces are used to mark the territory, but also to signal social status, and gain reproductive synchrony (Peters & Mech 1975, Martin et al. 2010). The territorial markings prevent possible aggressive confrontations at the same time as the markings help flock members to keep in contact with each other. The markings are placed along the borders of the territory, preferably at a cross-road and at mountain tops, where the scent has a good spread (Barja et al. 2004). Lone wolves, without a territory and a mate, keep a low profile and do not mark much, particularly not at the cross-roads where the territorial wolves mark. The individuals that mark the most belong to newly formed pairs. A particularly strong signal is the so-called "double marking" (Rothman & Mech 1979). This is when the wolf urinates both by squatting and by leg-raising. This is mostly done by the breeding alpha couple, and they are also the ones who are most interested in examining the markings. Rothman and Mech suggest that the double marking may have an intimidating effect on subordinate individuals as well a a positive effect for the breeding couple to be synchronized. In general, wolves are clearly stimulated by markings from others to mark over the earlier scent. Also dogs tend to mark more when there are already scents from others. In an experiment with displaced urine, it was found that male dogs urinated more often on markings from other males than on markings from females (Bekoff 2001).

The olfactory communicative system develops over time. In a study of faeces collected in five packs of Iberian wolves during a period of six months, Martin et al. (2010) found differences between adults and pups samples. The adults had a higher number of different compounds, but also a more specific modification in aromatic and fatty acids. The lack of these in the pups indicated that the anal-sac glands are not developed in young pups.

6.7 Developmental patterns in intra-specific communication

Wolves and dogs are born semi-altricial, i.e. they are deaf and blind but they have some capacities, for example crawling and moving the head. They crawl towards the warmth of the mother and move the head in circles to find a nipple to nurse. The social development is divided into the following phases:

- *Phase 1. The neonatal phase* (0–14 days). During this period the pups spend their time near the mother and siblings. They whine and yelp if they experience cold, hunger or pain.
- *Phase 2. The transition period* (14–28 days). Around the age of 14 days the eyes open and shortly afterwards also the ear-channels open. This is a period when the social development can be observed. The first barks are reported at 18–21 days, at the same time as the pups initiate play by tail wagging and play postures.
- *Phase 3. The socialization phase* (4–8 weeks). Social play increase in intensity. The play usually involves racing and biting and the pups shift roles from dominant to nondominant. Size and strength do not matter in the play situation, the smallest pup may play the role of dominant and fight down a stronger animal. During this phase differences between wolves and dogs emerge. Wolves use more facial expressions whereas dogs use more vocalizations (Feddersen-Petersen 1991). Around the end of this phase the wolf pups will leave the den and stay at a rendez-vous place while the adults are away hunting. At the same age, domestic dogs leave their mother and littermates to move in with a human family for a second socialization period.
- *Phase 4. Juvenile phase* (8 weeks – 1 year). Wolf pups start accompanying adults on hunts and find their way back to the rendez-vous by themselves. They start howling around the age of 3 months and also the mothers can be observed howling again. The mother refrains from howling when the pups are small, probably to protect the pups by not disclosing the den until the pups are more indepedent (Joslin 1967). During this phase dogs will interact more and more with humans.

As is evident from the passages above, the social development of the wolf and of the dog is similar. It is also clear that the human influence comes in at exactly the right timing: when the young wolf is ready to go out into the world, the young dog is met by the human family.

As mentioned earlier, the differences between different dog breeds is striking. Some have hair over the eyes, some wrinkled faces etc. It seems plausible that dogs will learn to interact with different breeds. But how do the differences affect intraspecific communication in puppies with no earlier experience? Kerswell and colleagues (Kerswell et al. 2010) studied puppies from different breeds interacting during a "Free play" situation in a "Puppy Socialisation Class" in Melbourne, Australia. The pups were between 8 and 20 weeks of age, belonging to 30 different breeds. They interacted for 10–45 minutes, in groups of 2 to 10 dogs. The interactions were video recorded and analysed according to behaviours that occurred during the play, like biting, pawing, vocalisations, pouncing, turn head away, etc. The puppies were also classified in types according to morphological structure (coat, ears, eyes, snout, etc.). It turned out that very few differences in social behaviour was correlated with differences in physiology. Two features stood out as important variables causing differences; snouth length and eyes covered by hair. Puppies with short snouts were more likely to sniff the head of the other pup, and they were also more likely to be bitten on their head, possibly as a reaction to their sniffing close to the head. Puppies with eyes partly or totally covered often turned their heads away from the other dogs, and they were more often bitten on the body. These reactions were independent of the breed of the other dogs, so even dogs with short snouts bit other short-snouted dogs over the head. Both snout length and eye cover are important to communicative behaviour. Dogs with long snouts have better olfactory skills, so they have no need of coming close in order to sniff. The eyes are important in communication, as dogs have been found to be able to follow the attention of other dogs by looking at their gaze (Horowitz 2009). If the eyes are covered, it is harder to know what the dog is looking at.

6.8 The dog in the human family – learning to communicate with another species

Dogs have their natural group in the human family. The humans provide food, shelter and company. During the years of domestication, dogs have become integrated as members of the human group. There are indications that dogs prefer humans to dogs already when they are young puppies. When young dogs and wolves were given a choice between staying with a human sitting on the floor and with a dog lying close by, dogs chose the human, whereas even wolves that had been socialized with humans, chose the dog (Miklósi 2007a, b, Gácsi et al. 2005).

6.8.1 Dogs' understanding of humans

One reason for the smooth human-dog interaction we are familiar with is that dogs, as well as humans, constantly read the signals about emotional status from the other group members and react by emitting own signals. These signals may be vocal, visual, tactile and chemical. Acoustically, both humans and dogs make use of the so-called 'frequency code' used by many mammals (Morton 1977, Ohala 1984). According to this, low frequency tones are associated with dominance, and high frequency tones with uncertainty and submission. This makes it easy for dogs to recognize emotions in the human voice. Also the intonation patterns used in speech refer to the frequency code. Almost all human languages express questions and appeals by rising, and declaratives and imperatives by falling intonation. Sharing a code like this facilitates inter-specific communication – the dog will easily hear if the owner is angry.

Another acoustic dimension is the rhythmic patterns used in communication. Short repetitive notes elicit an alerted behaviour in many species and are used in many contexts (e.g. alarms, ambulance sirens). In an investigation of whistle signals used by shepherds to control herding dogs striking similarities between individual shepherds were found (McConnell & Baylis 1985). Although all shepherds claimed to have assigned acoustic structures to meanings in an arbitrary way, "Stop" signals were always characterized by long, continuous tones, whereas "Go" and "Fetch" signals were characterized by short, repeated notes. McConnell (1991) did a follow-up experiment on these results, hypothesizing that long tones would have better effect than short tones in the training of "Sit" in a dog, whereas short tones would be more effective for the command "Come". Half of the puppies in the experiment were trained to come after 4 short notes and sit after 1 long note, and half of the dogs were trained to come after 1 long note and to sit after 4 short notes. They were then retrained to do the opposite. The results show a clear advantage for short notes for the command "Come". By using a natural sign for alertness, such as short repetitive notes, the trainer gets a short-cut to the dog's comprehension.

What about catagorical perception? Can dogs understand segments of human speech as well as the intonation patterns? Results from investigations of dogs' ability to discriminate demonstrate that dogs are able to discriminate human speech sounds, just as chinchillas, nonhuman primates and birds (cf. 1.2.5). Fukuzawa et al. (2005) designed an experiment where dogs were exposed to commands for "Sit" and "Come". The commands were recorded in three versions of Sit: [chit], [sat] and [sik]; and three versions of Come: [tome], [ceme] and [cofe]. The results showed that for the command "Sit", the variant [sik] did not cause any problems for the dog, whereas the variant [chit] had a significantly lower response. One explanation for the lack of difference between sit and sik is that the sound is changed into a similar sound. They are both unvoiced stops, the only difference being the feature high; in pronouncing /k/

the body of the tongue is high, which is not the case for /t/. For /ch/ and /s/, the differences are larger. They are both unvoiced sounds, but /s/ is uninterrupted whereas /ch/ is interrupted. Another explanation could be the position of the sounds – the sounds that are changed in *sit* and *sik* are in final position, whereas /ch/ and /s/ are in initial position. More data are needed in order to state whether it is the position or the features that are most important for the dog's discrimination.

Dogs' understanding of humans includes the learning of human language, with vocabularies of 100–200 words. Examples from family dogs shows that dogs do not just learn words as whole units, but they are able to use the words in different syntactic combinations, e.g. "bring X", "take X to Y", "put X in a box". They seem to acquire the words by the same mechanism of fast mapping, i.e. acquisition of a word by just a few exposures, as human children do (Kaminski et al. 2004).

When it comes to visual communication, the literature abounds of studies showing how well dogs understand humans. Dogs are inclined to look at human faces and they relate to human pointing behaviour, either by bodily direction, gaze, hand, arm or leg, even across the body (Miklósi et al. 2003). The only context that seem difficult to the dog is when the pointing is done with just finger and not arm (cf. Chapter 4 for pointing in human cultures). Showing a direction by some bodily orientation is probably an underlying universal in cooperative animals and it is also part of wolf and dog behaviour. Even if there are conflicting signals, smell and human pointing, the dog chooses to rely on the human pointing (Szetei et al. 2003).

6.8.2 Humans' understanding of dogs

Humans seem to understand dogs quite well. The shared frequency code makes it as easy for humans to understand emotions in dog vocalizations as it is for dogs to understand human voices. There is experimental evidence that human listeners are able to interpret the dogs' emotions and to rate the contents aggressiveness, fearfulness, despair, playfulness and happiness, on the basis of barks (Yin & McCowan 2004, Pongáscz et al. 2005).

Also many visual signals are shared and easy to understand. Humans understand dogs indicating something by means of body position, gaze and gaze alternation. Both dogs and humans make themselves smaller in order not to intimidate, and try to give the impression of a larger body in order to threaten. Some behaviours, like touching of the chin as appeasement signal, are typical to the dog, but have gained understanding among humans.

6.8.3 Not interactional synchrony – but accommodation

The interactional synchrony found in many social animals: primates, birds, sheep and horses does not seem to be part of Canine behaviour, the exception being howling concerts in wolf packs. Studies looking for imitation have found that dogs are able to learn by imitation (Range et al. 2011, Miller et al. 2009) but the immediate imitation which is automatic and probably the result of mirror neurons has not been observed. Instead, some kind of delayed imitation seems to take place. For example, after being exposed to how a door can be opened by pushing with the head or with the hand, dogs were able to learn to do the same way as the human instructor (Range et al. 2011).

However, there is a lot of accommodative behaviour found in human-dog interaction. Humans and dogs have developed means to modify their behaviour to meet the other half-way. Dogs have a large learning capacity and can adjust to human needs in different situations. Studies of guide dogs give specifically good insights in this ability. Dogs guiding the blind have to take into account and avoid any obstacles, judging their height and width against the size of the owner. This is a fundamentally co-operative event, where the human has initiative about when to start and the direction where to go, whereas the dog is responsible of obstacles on the way and decides when to step aside or slow down in order to avoid collisions (Naderi et al. 2001). When the human owner is blind, the dog cannot get attention by gaze alternation, but has to find another way. Gaunet (2008) found that some guide dogs used loud mouth licks in order to attract the attention.

Also humans accommodate to dogs, and display behaviour typical to dogs. In playing with a dog, humans use actions such as "pat floor", "play bow", "forward lunge" together with vocalizations in order to solicit play. In a study on 21 dog owners, the "bow" and "lunge" both gave the best result (Rooney et al. 2001).

Another kind of modification, not really accommodating, is the special way of speaking many people use when they address a dog. Studies of human communicating with dogs have shown that people talk to dogs almost like they talk to infants (Baby Talk, cf. 3.2.1). In the case of Baby Talk, in has been claimed that the adult tries to accommodate to the voice of the infant, at the same time as it is a way to teach language. In the case of Pet Talk (Burnham et al. 2002), neither explanation is probable – humans do not try to sound like the dog, and they are not teaching the dog to talk. Instead, it may be the emotional side of the modifications that are important. The box below summarizes the features that are characteristic to Baby Talk and to Pet Talk.

The features that are typical are high pitch, exaggerated intonation, short utterances, many imperatives, small vocabulary, and many repetitions. These features are used to attract attention. Other characteristics, such as hyperarticulation of certain speech sounds (as [i] in 'sheep', [u] in 'shoe' and [a] in 'shark'), use of questions and expansions, can be seen as ways to make speech clearer and thus as ways to teach language. The hyperarticulation makes it easier for the child to find the language-specific pronunciations, the questions and expansions are there to get the child involved in the interaction. Interestingly, and as could be expected, only the attention attractors are used in Pet Talk, and not the language teaching devices. These would be quite useless.

Box 6.3 Features of Baby Talk and Pet Talk (after Ferguson 1977, Kuhl et al. 1997, Burnham et al. 2002)

FEATURES	BABY TALK	PET TALK
High pitch	yes	yes
Exaggerating intonation	yes	yes
Hyperarticulated vowels	yes	no
low MLU	yes	yes
Present tense	yes	yes
Imperatives	yes	yes
Questions	yes	no
Limited vocabulary	yes	yes
Few pronouns	yes	?
Repetitions	yes	yes
Expansions	yes	no

6.9 Summary

This chapter has given an overview of communication in the domestic dog, and it has also presented some research on the ancestor of the dog, the wolf. Sometimes the behaviours are easier to study in wolves, because of the enormous scope of variation between different dog breeds. Dogs differ from other animals in that they are socialized both with their own species and with humans. Therefore we know a lot about the social functions of dog communication and our inter-specific communication runs smoothly. The long period of co-habitation has probably had an impact on dogs' understanding of humans, as well as humans' understanding of dogs. The use of referential communication in dogs is similar to, or superior to, that of nonhuman primates, in that they have no difficulties in learning to communicate human referential means, such as pointing and words referring to external phenomena. There are, however, no evidence of use of referential function in intraspecific communication.

6.10 Suggested readings

Jensen, P. (Ed.). (2007). The behavioural biology of dogs. Wallington, Oxfordshire: CABI International Press.

This book can be recommended to students of communication and dog lovers alike. It consists of four parts; evolution and domestication of the dog; behaviour and learning; the dog among humans; and behavioural problems. The different chapters explore aspects of social behaviour, social cognition, learning and human-animal interactions. Studies of the behaviours in particular settings, such as free-ranging dogs and working dogs demonstrate the extent of variation and complexity in dog behaviours.

Mech, L. D. & Boitani, L. (Eds.). (2003). Wolves. Behavior, Ecology, and Conservation. Chicago: The University of Chicago Press.
This is the standard work on wolves, edited by the leading researchers in the field. It brings together research on evolution, ecology, social behaviour, communication, hunting, interactions with other species and conservation issues. Examples are taken from all over the world, e.g. from the US, Canada, Italy, Israel and India. The book is richly illustrated with colour pictures.

Coppinger, R. & Coppinger, L. (2001). Dogs. A New Understanding of Canine Origin, Behavior, and Evolution. Chicago: University of Chicago press.
The authors have practical as well as theoretical experiences with dog. The book discusses the behaviours of different types of breeds – such as herding dogs and sled-pulling dogs – and how these different breeds came to be. A large part of the book deals with the relationship between dogs and humans.

Chapter 7

Communication in birds

7.1   Introduction

As mentioned earlier, comparisons between humans and animals are motivated by either a genetic relationship between the species, or by a similarity in behaviour. The reason for including a chapter on communication in birds in this volume lies in the similarities – in this case not in the social structure, but in the ability for *vocal learning*. The vocal communication in birds is usually divided into two types, songs and calls. Songs are long, elaborated vocalizations with melodic tonal contours, predominantly produced by male passerine songbirds, and mostly during the breeding season. They often happen spontaneously (i.e. are not triggered by a particular event in the environment) and have a typical diurnal rhythm. Calls, on the other hand, are shorter and simpler sounds that may be produced by both males and females all year round. Calls are often reactions to (changes in) the environment and they express a range of different functions, such as alarm, flight or discovery of a food source (Catchpole & Slater 1995). However, calls are also used for individual recognition and are therefore subjected to variation at the same time as they serve as referential signals. In this chapter we will present studies on referential and social functions of bird communication. First a few words about the enormous size of the group of species joined together under the name of birds, or the Aves class.

7.1.1   The Aves class

There are approximately 9,000 recognized species of birds, divided into about 23 orders and 180 families. They range in size between the tiny hummingbird (*Archilochus colubris*; 10 cm, 3 g) and the large ostrich (*Struthio camelus*; 2.5 meters, 100–160 kg). A general characterization of birds includes the features bipedalism, feathers, wings, beak with no teeth, and hollow bones. Most birds can fly, but not all, e.g. penguins (the family *Speniscidae*), ostrich, and kiwis (the family *Apterygidae*). Many have monocular vision, with the eyes placed on the sides of the head – this makes it easier to discover predators (predator species like owls have different placements).

More than half of the bird species (5712 out of 9672 according to Sibley & Monroe 1990) belong to the order of Passeriformes (also called the perching birds). Of these, around 4500 are so called true songbirds, or *Oscines*, for example finches (*Fringillidae*) and wrens (*Troglodytidae*). The rest are *Suboscines*, for example woodcreepers (*Dendrocolattidae*) and the New World flycatchers (*Tyranidae*). The Passeriformes are characterized by having four toes, three forward and one (the homologue of the human big toe) directed backwards, which helps perching on a branch, whereas many other birds have other types of feet, for example the parrots (*Psittaciformes*) which have two toes front and two back and the ducks and geese (*Anseriformes*) which have webbed feet.

Most research on communication in birds has focused on song learning in the Oscines. It is particularly songbirds in temperate zones which have attracted the interest. These have a short mating period during which their singing makes a strong impression on the environment. Oscines (true songbirds) learn their song by using another bird as a model, whereas suboscines are able to develop their songs even if they are brought up in isolation. Among those who learn from a model, the learning requirements differ; some species need a live model, others can do with tape recordings, some need their father to teach them, others can have other adult teachers, some imitate with exactness, and others improvise a lot. However, as Bretagnolle (1996:160) points out, songbirds ought not to be treated as the most typical member of the class *Aves*. There are many birds that do not learn vocalizations, many where females do

Box 7.1 The family tree of the Aves class – simplified

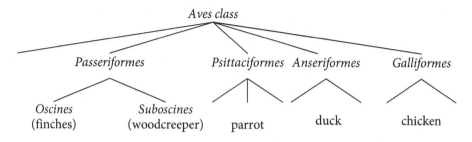

not select a partner for his song, and many where visual signals are as important as acoustic ones.

Considering the large number of different species, bird communication cannot be treated as a uniform behaviour. This chapter is limited to some selected species as examples of communicative behaviour. The presentation will have two main foci; vocal learning in *Passerines* (songbirds) and the repertoire of calls with identifiable functions in *Galliformes* (for example domestic hen).

7.1.2   Social life of birds

As in all living species, the communicative needs are associated with the demands of social life, and with the wide range of options in birds, this points toward a lot of variation. There is a continuum in terms of group size from birds that only meet to breed, to long-term pairs, team-hunting predators, family groups, harems with many females, to large clans, and migration groups. Like primate groups, many birds live a fission-fusion social organization – they split into small forage groups and then reunite in the main group.

In certain bird orders, for example *Galliformes* and *Psittaciformes*, the groups are socially structured and patterns of dominance are established in daily interactions (Bayly et al. 2006). Among these birds, it is typical to have many different kinds of calls, for example to inform members about dangers that may threaten the group, and food that may be shared. Many galliformes, living in seasonal harems with a rooster and some females, have large repertoires of differentiated calls with different meanings attached to them. In red jungle fowl (*Gallus gallus*) and domesticated hen (*Gallus gallus domesticus*), up to 18 differentiated call types have been identified (Collias 1987, Marler 2004b:135). The *Psittaciformes* (e.g. Grey Parrot *Psittacus erithacus* and budgerigar *Melopsittacus undulatus*), on the other hand, live in flocks that can be quite large, with several small stable flocks forming one large temporary group. The plasticity in their vocalizations allows for the acquisition of different calls for group identification as the bird matures and moves "from family group to juvenile crèche, to juvenile flock, to local nomadic flock, to resident breeding flock" (Farabaugh & Dooling 1996:107). Parrots are known for sentinel behaviour, i.e. while the group is foraging, there is a guard that watches out for dangers (Yamashita 1987).

Cooperative breeding, involving alloparents in the care of juveniles, occurs in around nine percent of all bird species (Cockburn 2006). There are three conditions that seem to favour development of cooperative breeding in birds. Cooperative breeding is more common in (a) birds with slow maturation and long lifetime; (b) birds that stay in the same area year-round; and (c) birds in environments with variable food access, for example with unpredictable weather conditions. During periods of hard weather it may be hard to find provisions for the young without help from alloparents. Cooperative breeding is particularly common in the *Corvidae* family (Ekman & Ericson 2006). Corvids (e.g. common ravens, *Corvus corax*; and jackdaws, *Corvus monedula*) are known for complex behaviour, for example cooperative hunting, also rich repertoires of vocalizations and in some cases also for an ability to imitate human speech. A further skill is an ability to learn from others, and a relationship between learning and cooperative breeding has been suggested: "It leads me to inquire whether there is some interaction between a deep history of cooperative breeding and offspring that grow up to be especially good at learning from others and manipulating their physical as well as social environments" (Hrdy 2009: 198; cf. also Chapter 8).

7.1.3   Observation techniques

The chorus of birds at dawn and dusk is impressive, and it is difficult not to notice their acoustic expressions. But for description and analysis it is necessary to make recordings. Modern technology offers a wide range of possibilities to do empirical research on bird communication, for example, audio- and video-recordings, playback of vocalization to individuals in their natural environment and computer-assisted analyses of sounds. Before the invention of sound spectrographs, the only way to render bird vocalizations for analysis was by different types of transcriptions, either in form of musical notes, or, more commonly by using the ordinary alphabet. This is still the method used, for example in field guides. A quick glance in a bird guide shows, however, that the notations are onomatopoetic approximations and the descriptions are determined by the perception of the transcribers. Comparisons between descriptions in bird field guides reveal language-specific differences. To take some examples, the sound of the skylark (*Alauda arvensis*) is described as "chirrup" in Australian field guides, and as "tchriup" in the Swedish guide. The sanderling (*Calidris alba*) says "twick-twick" in Australia but "plytt" in Sweden. The laughing turtle-dove (*Streptopelia senegalensis*) says "cooo-wook-coo" in Australia, but "poo po poO-pOo-ho" in Sweden. Finally, the green finch (*Carduelis chloris*) says "swee-eee-e" in Australia, but "djyrrrypp" or "dschväsch" in Sweden. Such descriptions may be helpful in field use but are, of course, awkward to use for comparative studies.

New technology provides more objective and consistent representations. In a spectrogram the composition of a sound with respect to time, frequency and intensity is showed in great detail. With the invention of the sound spectrographs in the 1950s came a completely new way to capture sounds in order to analyse and describe them. One of the pioneers in this area, the biologist Peter Marler, remembers the "good old days":

> Until about 1950, everyone interested in birdsong had no choice but to work by ear. Only when the sound spectrograph became available was it possible, for the first time, to grapple objectively with the daunting variability of birdsong, and to specify its structure precisely. Almost immediately a multitude of new issues became accessible for scientific scrutiny and experimentation. (Marler 2004a: 1)

One of the issues that could now be dealt with was field identification of species. Thanks to the new technique it also became possible to identify universal and variational features and describe variability within and between individuals of the same species, for example in terms of geographical variation. It takes some training to interpret spectrograms, but once this is learned it is possible to interpret spectrograms just as well as orthography, and this method has the potential to make field manuals more precise.

Playback is an extremely useful tool in finding out about the meanings of bird vocalizations. By playing back a sound to a bird it is possible to decide whether the sound in question is communicative or not, and by systematically adjusting the features in the sound track it is possible to distinguish exactly which features of the sound are significant (Evans 1991). The interactive playback gives even more information about behaviour. It consists of several steps. When the focal birds respond to the first playback a second sound is played, with a segment that fits the song sequence of the bird (Dabelsteen & McGregor 1996).

Most observations on song learning have taken place in laboratories, which may not be the optimal way of finding out natural behaviour. As was discussed in Chapter 2, it may be problematic to generalize results from animals in captivity to their natural behaviour in the wild. "I am increasingly convinced that laboratory studies can at most show only what a bird is capable of doing in an environment never before encountered in the species' evolutionary history" (Kroodsma 1996: 4).

The acoustic and the visual channels are the ones most investigated in bird research and which have the most advanced methodology. Tactile communication can be observed in mutual preening, but it has not been systematically investigated. There are very few studies on chemical communication, showing that it may play a role in some species, for example mallard ducks (*Anas platyrhynchos*; Jacob et al. 1979).

7.2 Functions – why do birds communicate?

Birds need to communicate to find a partner, keep away competitors, bring up a family, warn against predators, inform about food sources and to keep contact during migration and other movements (cf. Collias 1960). These functions are fulfilled in different channels: acoustic, visual, tactile and chemical.

The most striking communicative behaviours in birds are the vocalizations – and among the vocalizations, the songs of the song birds are particularly conspicous. Most songs are performed when males are looking for partners, and the expression can be interpreted as an announcement like "Listen, I am a chaffinch and this is my territory, I want to raise a family here so females are welcome but males should beware". The song in the passerines and the crowing in the galliformes are both stimulated by the male hormone testosterone, they have clear sexual connotations, serving at the same time to attract a female and to defend a territory. The biological value of the vocalization among galliformes is shown by the fact that the males that crow the most also have the largest reproduction rate (Wilson et al. 2008).

7.2.1 Alignment of songs – counter-singing and duetting

Songs are sometimes monologues, but they can also come out in form of dialogues, with interaction and alignment between individuals. In species with large song repertoires, males often engage in what is called matched counter-singing, or song sharing, for example by replying to a neighbour's song with a similar song, or, if his song is well-known, even start singing the next song in the sequence, thereby taking over the initiative (Catchpole & Slater 1995, Kroodsma 2004). To counter the song of a neighbour in this way can be considered a challenge, and evoke a response of aggression (Burt & Vehrencamp 2005, Peake et al. 2005), but it has also been noted to result in social success in terms of increased chances to attract a mate (Poesel et al. 2012). Some birds that are known to exhibit this behaviour are the great tit (*Parus major*; McGregor et al. 1992), the song sparrow (*Melospiza melodia*; Beecher 1996), the white-crowned sparrow (*Zonotrichia leucophrys*; Poesel et al. 2012), the common nightingale (*Luscinia megarhynchos*; Todt & Hultsch 1996), the black-capped chickadee (*Poecile atricapillus*; Hailman & Ficken 1996), and the golden whistler (*Pachycephala pectoralis*; van Dongen 2006).

Counter-singing interactions are common between neighbours. To get to know the neighbour, and his song repertoire, is a good strategy since it is an advantage to have familiar neighbours and well established territories, the so called 'dear enemy' effect (Catchpole & Slater 1995). To know the neighbour's song repertoire gives the male full control over the interaction and he can choose whether to escalate or de-escalate. Beecher and Brenowitz (2005) report that the song sparrow has a number of choices to make in order to regulate the interaction, and each choice decides the possibilities for the next.

Also in the golden whistler the different choices decide what will follow in the dispute. van Dongen (2006) used interactive playback to simulate counter-singing and found that the first forms to be used is song-type switching (i.e. switching between songs), and song matching (i.e. match the song to the playback song, or parts of it), but if the intruder comes closer the golden whistler male starts overlapping the song from the intruder. It can be assumed that song overlapping is the most aggressive since it results in the intruder's song not being distinguishable.

Counter-singing is an example of an interaction between males and it can be used in territorial disputes. Another kind of cooperative singing, which maybe is more similar to human alignment or synchronization, is the duetting between male and female (Collins 2004:78, Farabaugh 1982). One difference between duets and counter-singing in males is that the duets have a more precise timing, with a fair amount of overlapping between bouts or phases. If the male and female sing alternately the term antiphonal singing has sometimes been used. "Indeed, while bouts may overlap, the sounds themselves may not do so, the birds fitting their sounds together so precisely that it is hard to believe that more than one individual is involved" (Catchpole & Slater 1995:174). Duets are acoustic, but do not have to be vocal, since they can also involve drumming of woodpeckers. Their functions are assumed to be the pair-bonding, at the same time as it is territory-claiming, since it indicates a stable relationship and signals to other males to keep away. Duetting is most commonly found in the tropics, where the birds do not migrate, but have relatively constant territories the year round. Other features that are typical to duetting birds are that females also sing and that there are permanent pair-bonds (Farabaugh 1982, Morton 1996, Marshall-Ball et al. 2006).

7.2.2 Referential function in birds – the calls of the domestic fowl

Songs are typical expressions for advertising, mating, territorial defence, which constitutes the reproductive function. However, birds may also communicate about an external referent, that is, use the referential function (see Chapter 1). In a series of experiments on communication in the domestic fowl (*Gallus gallus domesticus*) it was demonstrated that domestic fowl produces calls that convey precise information about predators, as well as about accessible food resources. For predator warnings, there are different calls for enemy from the ground and enemy from the air. When these calls were played back to hens in experimental setting, totally different effects were found. The reaction to a ground alarm call is to stand up and scan the environment to detect an enemy, whereas the reaction to an aerial alarm call is to look upwards, crouch down with head and tail lowered (Evans et al. 1993, Evans & Marler 1991, 1995, 1995, Marler et al. 1993). To investigate the exact meaning of an alarm, whether it meant "something overhead" or "an aerial predator" one of the experiments included video images, where animated sequences of ground predator (a racoon) and an aerial predator (a hawk) were displayed. The images were placed in different places, at ground level and over the cage with the male chicken (i.e. racoon on the ground and racoon in the air and hawk on the ground and hawk in the air). When the predators were placed at unusual positions there were fewer calls produced. The hawk elicited mostly aerial calls no matter where it was placed (even though there were some attempts at ground calls when it was placed on the ground). The responses to the racoon were mixed: while it elicited only ground calls when it was placed on the ground, there were more aerial calls than ground calls when it was positioned overhead. A possible explanation for this apparent confusion is that some contextual information was missed in the ground position, since the image was constructed in such a way as to have a neutral background, and the chicken probably expects more background information on the ground (e.g. bushes) than from the silhouettes in the air (Evans & Marler 1995).

Another call with a referential function is the food call (or resource-recruitment signals, Bradbury & Vehrenkamp 1998). These calls are produced when food is discovered, alerting conspecifics that there is a forage source to be shared. In the domestic fowl, the response to a food call is a downward movement and ground inspection, (Marler et al. 1986a, b, Evans & Evans 1999, 2007). The use of food calls is interesting from many aspects. They are used by many bird species, in addition to the domestic hen, for example, herring gulls (*Larus argentatus*; Morton 1982), ravens (*Corvus corax*; Heinrich 1988), house sparrow (*Passer domesticus*; Elgar 1986), and cliff swallows (*Hirundo pyrrhonota*; Brown et al. 1991). The fact that the calls are used much more when there is an audience at hand, than when the caller is alone (e.g. Evans & Marler 1991, 1994, 1995), shows that the calls are not just automated reflexes, but directed towards a potential listener. There is an audience effect, i.e. the audience influences the behaviour. But why attract others to a food source you can eat all by yourself? One suggestion is that bringing more birds together diminishes the risk for the caller to be attacked by a predator. However, this explanation does not fit the results from domestic fowls, since the males are known to call *more* when there is a female around than when there is a male (Evans & Marler 1995). This suggests multifunctionality – a food call can function as a courting call in certain contexts.

The studies of audience effects have broadened the scope of communicative functions found in birds. If not only humans, but also animals are able to consider the effects of their communicative behaviour, and help others in defence and foraging, some kind of reciprocal altruistic behaviour could even be assigned to them, so that they help others in order to get help when they need it. However, as mentioned earlier, the whole discussion of communicative functions in animals is flawed by the fact that we don't really know what goes on in the minds of the individuals, since all we have is their behaviour and our interpretations.

7.3 Acoustic communication

Acoustic signals play an important role in birds' communicative behaviour. Discovering a bird during a walk in the field or the forest is in fact more likely to take place by hearing it than by seeing it. The acoustic signals do not have to be vocal, but can also be made by other means. The tapping on wood by the woodpecker, for example, is a well-known acoustic signal, with the same potential for advertising and defence as the song in the songbird. A less known behaviour is the sound production made by the specially developed wing feathers in club-winged manakins (*Machaeropterus deliciosus*). These birds produce sounds by quickly flipping the wings above the back, in a way more similar to insects than to other birds (Bostwick & Prum 2005). Songbirds detect a lot of information about others by their song performance. For example, females seem to prefer males with rich song repertoires (Lambrechts 1992, Hasselquist 1994, Hasselquist et al. 1996, Nowicki et al. 2000, Nolan & Hill 2004), possibly from learning (Lauay et al. 2004). A bird with a large song repertoire may signal a better physical condition (Kipper et al. 2006), higher capacity to feed nestlings (Buchanan et al. 1999), higher survival and fighting ability, which in turn have a positive influence on the number of surviving offspring. This implies that the song serves as a marker of the male's value as parent. Furthermore, Hasselquist et al. (1996) found that, if they had extra-pair matings, females of the polygynous great reed warbler (*Acrocephalus arundinaceus*) preferred a male with larger song repertoire. Even though males with large song repertoires did not provide more paternal care, the offspring had larger survival rate, which implies an indirect genetic benefit of choosing a male with rich vocal performance.

Acoustic details in song may also provide information about the distance to the individual. This plays an important role to territory holding birds, and helps them to spread and to avoid unnecessary confrontations. Experiments with playback show there are features of the sound that make birds able to pinpoint the exact distance, particularly when the song or call type is familiar (Naguib 1996, Naguib & Wiley 2001, Nelson 2000).

A lot of research on acoustic communication has dealt with describing the tones. Generally, high tones are typical to small birds and low tones to larger birds. One example of a low sound is the vocalization of the large cassowaries (*Casuarius casuarius*) in Papua New Guinea. These birds live solitary lives, and when the males produce sounds to attract females during the breeding season, they use frequencies down to 20–30 Hz (Mack & Jones 2003). This is an optimal way to reach through the dense rainforest. A small bird, for example the blue-throated hummingbird (*Lampornins clemenciae*), on the other hand, may produce high ultrasound frequencies up to around 30 000 Hz. However, in a study comparing production and comprehension, Pytte et al. (2004) found that the highest tones were not perceived by the hummingbird, and thus, they are not used for intra-specific communication, but may have other functions.

Interestingly, birds do not produce sounds in the same way as mammals do. In the human case it is the larynx that produces many of the sounds. Birds also have a larynx, but for birds it has a different function, namely to regulate the airflow during the song. In most birds, it is the syrinx, which produces sound. The syrinx is situated further down in the trachea, between the primary bronchi and the heart. The song is created by air flowing over thin membranes that start to vibrate. The largest difference between bird and human vocal production is, firstly that the syrinx has two different sources (at least in some birds), one in each bronchus, and secondly, that songbirds do not modify the sound by using mouth or tongue. Instead, they use the beak, opening-closing it to different degrees, to alter the sounds they produce. Moreover, many birds also lengthen and shorten their trachea, in combination with the pressure changes of the syrinx membranes, to produce different sounds. The variation in bird song is also a result of the activity of muscles in the syrinx, and the fact that sounds are created from both sides of the syrinx simultaneously. Songbirds may use only one side, or both sides at the same time. It is possible to identify which side of the syrinx is responsible to which part of the song, for example in the catbird (*Dumetella carolinensis*) the syllables in the complex song are produced at different sides (Suthers 1990). The double effects of the syrinx allow the bird to take mini-breaths during singing. There is also another way to optimize the use of breathing. Certain high notes in the zebra finch (*Taeniopygia guttata*) are produced on both inhalation and exhalation (Leadbeater et al. 2005). A functional use of the two sides of the syrinx has been discovered in the emperor penguin (*Aptenodytes forsteri*). The call for individual recognition used by the penguins consists of two series with different fundamental frequencies. By using playback experiments, it was possible to show that a call with only one of the frequencies did not result in individual recognition, whereas a call with both frequencies was recognized (Aubin et al. 2000, see also Sturdy & Mooney 2000).

Research on the neurological correlates to bird vocalizations has revealed that there are specific brain mechanisms for the recognition and storage of songs in songbirds. These specific neural circuits are developed in birds that learn their songs, and have a need to store the model songs and compare them to their own vocalizations. (Nottebohm 1999, Whaling et al. 1997). We will return to this issue in the section on learning below (Section 7.4).

7.3.1 Variation across and within species

The amount of singing heard from songbirds varies considerably between species. Some species sing during the whole breeding season, whereas others sing less and less or terminate directly after breeding. Among migrating passerines in temperate zones it is common to meet only for the breeding period, during which males devote their energy to singing highly sophisticated and variable songs. After that period there is no need to attract females and defend the territory by singing. Tropical passerines, on the other hand, often live in monogamous pairs and defend their territories year-round. In these species both males and females may sing and duet singing is a common phenomenon (Morton 1996, Marshall-Ball et al. 2006).

The variation can be related to the function of the song. If the function is predominantly to attract a partner the song period is short and intense and stops after pair formation, whereas if the main function is to defend a territory and ward off rivals, the song period is longer. In the latter case the characteristics of the song is different: it is simpler and interspersed with pauses when the singer stops to listen for reactions from expected intruders. The great reed warbler (*Acrocephalus arundinaceus*) sings an intense, loud, complex song, called the long song, to attract females. However, immediately when he forms a pair bond with a female (i.e. when a new female that has entered his territory accepts to copulate with the male) he switches to singing a less conspicuous, sparsely produced, simple song called the short song (Hasselquist et al. 1996). The male will sing the short song for about five days until the female starts laying eggs. Then the male resumes the long song to attract a new female.

One of the characteristics of singing is that it has a regular daily rhythm, being most intense at dawn and dusk. Different species have their specific place in the morning scheme: insect-eaters and worm-eaters start early while waiting for the insects to come out, then they have to search for food and stop singing. Then it is time for seed-eaters. The song starts with scattered strophes, and then it increases in intensity. The singing continues during the whole morning, but the top intensity lasts only for about half an hour. Around midday the song weakens and stops, and then there is an increase during the afternoon and a new peak in the evening hours. It has been observed that if humans provide birds with food, these birds sing more than others and they also have a higher mating success (e.g. Davies & Lundberg 1984, Cuthill & MacDonald 1990). The risks with singing, namely that singing birds are more exposed to predators and aggressive rivals, and that they have to devote some time and energy also to food exploration, are outweighed by the benefits. However, it is important not to waste more energy than necessary. Like many other animals (including humans), birds are able to make vocal adjustments in accordance with environmental needs. One example is to modify the vocal amplitude by increasing the intensity when there is a noisy background, and decrease it when it is calm, the so-called Lombard effect (after Lombard 1911). This has been found for example in zebra finches (*Taeniopygia guttata*; Cynx et al. 1998) and in blue-throated hummingbirds (*Lampornins clemenciae*; Pytte et al. 2003). For the great tit (*Parus major*) Hunter and Krebs (1979) found that birds recorded in the woodlands of Oxfordshire were more similar to birds in a similar environment in Iran, 5000 km away, than to the birds in New Forest, less than 100 km away. Studies in urban areas have got similar results: a higher pitch in a noisy surrounding, than when singing in the quiet areas of the city (Slabbekoorn & Peet 2003, Katti & Warren 2004). A result of a fine-tuning to the environment is that it preserves a constant broadcast range, which is valuable in territorial birds, since unnecessary loud singing may challenge territory holders far away. Slabbekoorn and den Boer-Visser (2006) recorded the song of the *Parus major* in ten cities from seven countries (Belgium, the Czech Republic, England, France, Germany, the Netherlands and Luxembourg) during the month of April 2002 and 2003. The city recordings took place in the core parts of the cities, e.g. downtown business centre, with most buildings at least four-stories high and less than 15% vegetation. The forests chosen were old-growth deciduous or mixed forests. In total, songs from 465 birds were recorded: 213 urban individuals and 252 forest individuals. The results showed similarities between the urban birds in the different parts of Europe. Urban features had shorter duration between songs, shorter duration of the first note, more notes, and higher frequency – in other words, it gave a hurried impression.

The auditory impression of song is often combined with a visual impression, since many songbirds place themselves in a conspicuous place like a top of a tree and are easy to observe when they sing (e.g. wrens). Some birds sing during flight (e.g. sky larks), and others (e.g. tits and nightingales) keep to shrubs when singing.

7.3.2   Structural aspects of song

The species-specific song in songbirds functions as a trigger for mating and defence, and it has to be constructed in a way to be precise and easy to recognize. Some song parameters that have been discussed and empirically tested for their relevance for species recognition are: frequency range, frequency changes between elements, intervals between elements, structure of individual elements, and the order of elements, or syntax (Becker 1982). Many species seem to be sensitive to a combination of parameters.

Box 7.2 shows which parameters are used in 12 oscine species. However, considering that there are many question marks, about 4500 more songbirds, and probably more parameters that have not been examined, the list should be treated as a suggestion of what is possible, not what is actually used.

Paradoxically, although each songbird species is associated with a characteristic species-specific song, most birds have several versions of their song. These versions form the individual bird's song repertoire. A song repertoire can thus be described as consisting of several songs, which differ in composition of recurring patterns. A wide range of proposals as to which units to use are given in the literature, for example: notes, elements, syllables, syllable types, phrases, themes, phases, song figure, songs, and song types. A three-level hierarchy of units has been proposed (consisting of elements, phrases and songs: von Dongen 2006) as well as a four-level hierarchy (with

Box 7.2 Parameters for distinguishing species-specific song. A (+) indicates important, (−) indicates not important, (?) indicates that the significance is unclear (after Becker 1982: 224)

SPECIES	FREQUENCY RANGE	FREQUENCY CHANGE	STRUCTURE OF ELEMENTS	INTERVALS BETWEEN ELEMENTS	SYNTAX: ORDER OF ELEMENTS
White-throated sparrow (*Zonotricia albicollis*)	+	+	+	+	?
Wood lark (*Lullula arborea*)	+	?	+	+	+
Ovenbird (*Seiurus aurocapillus*)	+	?	+	+	+
Willow Warbler (*Phylloscopus trochilus*)	+	?	+	+	+
Yellowhammer (*Emberiza citrinella*)	+	?	+	+	?
Indigo bunting (*Passerina cyanea*)	?	?	+	+	−
Robin (*Erithacus rubecula*)	+	+	−	−	+
Goldcrest (*Regulus regulus*)	+	+	−	−	+
Firecrest (*Regulus ignicapillus*)	+	−	+	+	−
Marsh Tit (*Parus palustris*)	?	−	+	+	−
Chiffchaff (*Phylloscopus collybita*)	+	−	+	−	−
Bonelli's Warbler (*Phylloscopus bonelli*)	+	−	+	−	−

elements, syllables, phrases and songs: Catchpole & Slater 1995). Using a three-level model, Todt (2004: 3) gives structural parallels in units of bird and human communication as is illustrated in the following:

Box 7.3 Structures in human and bird vocalizations

Birds		Humans
Singing (= bout of songs)	-------------------	Talking (= bout of sentences)
Songs	------ ------- ------	Sentences? Words?
Syllables	-- -- -- -- -- -- -- -- --	Syllables

The model gives a rough overview of commonly used units, and, as a thought experiment, we can expand it by adding intra-stage units such as elements (birds), phonemes, morphemes (humans), phrases (both birds and humans), clauses (humans). One difference between songs in a song repertoire is that units (elements, syllables, phrases) may be left out in the new song, another difference may be that the individual units are arranged in different orders. This can be compared to human syntax (see Chapter 3).

The song of the sedge warbler (*Acrocephalus schoenobaenus*) is famous for having a large repertoire of syllable types and therefore, has a great potential for syntax. Catchpole (1976) points out that : "a male sedge warbler may never repeat exactly the same sequence of elements twice during the course of his life. This is because the song is extremely long, and the 50 or so elements that a bird possesses follow each other in highly varied orders". In the song of another warbler, the great reed warbler (*Acrocephalus arundinaceus*), the number of syllables is lower but the principle is the same; units are put together in a large number of different combinations (Hasselquist et al. 1996).

In the song of the common nightingale (*Luscinia megarhynchos*), it is possible to discern a developmental hierarchy for the order of song units (Todt & Hultsch 1998). An adult nightingale may have a repertoire of 200 different song types at his disposal, with the units organized in various ways. By analysing the songs into smaller units, Todt and Hultsch managed to find systematic sequences of units that can be represented in flow chart, as each song is a sequence of decisions. Some units are only found in particular positions of the song, for example in initial position. In fact, many songs share the initial phase and differ only in the continuation. After the initial element, the choice of units is already constrained. The process is further facilitated by a tendency of some units to occur together in smaller or greater packages, like compounds. Adult birds use these packages in automatized procedures to facilitate retrieval in situations demanding rapid vocal responses. During early phases in development, juveniles sing incomplete songs with some units missing, and use only small unit packages. This developmental series reminds of the stages that have been observed in language acquisition, where processing procedures are acquired in a hierarchical order (see Chapter 3.5, Processability Theory).

7.4 Learning how to sing

How songbirds learn to sing has been a central theme in bird research since the pioneering work by Thorpe (e.g. Thorpe 1958) in the 1950's. The fact that on the one hand there is a stability of species-specific songs across time and geographical distribution and on the other hand dialectal variation has led to the hypothesis that the song is both innate and learned.

Experiments on birds raised in isolation show that they form songs that contain at least some characteristics of the conspecific song. This has led to the suggestion that songbirds have a blueprint (Thorpe 1958), an innate template (Marler 1970a, b), or perceptual constraints (DeVoogd & Székely 1998) with neurological correlates (Marler 1999, Whaling et al. 1997), that drive them towards species-typical sounds. This is in accordance with the observation that birds generally do not copy songs from other species in the environment, but keep to their conspecific songs. Furthermore, the songs do not change character over time, across generations, but the same typical characteristics remain. However, as in many cases of behavioural ontogeny, the development of vocalizations in birds is the result of an intricate interplay between innate predisposition and environmental experience. The proportions differ between species, so that some birds seem to have more innate vocalizations and learn very little, whereas others have larger potential for vocal learning.

7.4.1 Sensitive phases and developmental stages

Studies on development have shown that there are several phases or stages towards the target. There are two main phases in the acquisition of song. The first is when the young bird listens to the adult model (the sensory phase), the second is when the bird practises his own output (the sensorimotor phase). The exact timing of the first sensitive listening phase (sensory acquisition, White 2001) differs between species, but usually it is set before the age of 4 months. After hearing the song, the bird will memorize it during a silent period, during the winter. In spring, it is time for the next important phase. This is the sensorimotor learning, when birds practice singing, the syringeal musculature develops and the movements and opening of the beak are coordinated more and more with singing (Podos et al. 1995). During this period distinguishable stages can be heard: the subsong stage, the plastic song stage and the full, crystallized song. The first stage, the subsong, contains a number of highly variable vocalizations bearing little similarity to the song model that the birds have been exposed to. The plastic song is characterized by an overproduction, which gradually disappears after the interaction with other birds (Marler & Peters 1982). This process has been termed "selective attrition" and implies actively forgetting about certain songs. It is not known, however, whether this "forgetting" only takes place on the production level but also in perception (which it does in human infants). For some species the full-accomplished song is reached after some months, whereas other species (for example the starling and the nightingale) continue to develop the song during several years (Todt & Hultsch 1998, White 2001), even though the learning ability decreases with age (Eens et al. 1992). The general milestones in song development are summarized in Box 7.4.

Box 7.4 Milestones for early song development (after Hultsch & Todt 2004)

1. Subsong: soft sounds, high variability
2. Plastic song: precursors of adult structures, fragmentary and/or inverted
3. Song crystallization: the plasticity 'freezes', variability decreases, species-specific songs

The early stages can be regarded as stages of practice of motoric skills, just like the case with human infants' babbling. For birds, it is a matter of beak movements.

Some birds are close-ended learners and have to learn all details about their song during the sensitive phase whereas others are open-ended learners and are able to learn new songs for several years (e.g. starling, *Sturnus vulgaris*; Eens et al. 1992). There are also differences between birds that have one peak (e.g. zebra finch, *Taeniopygia guttata*) and those with two peaks with intervals between (e.g. swamp sparrow, *Melospiza georgiana*). Figure 7.1 illustrates the almost overlapping sensitive phase and production phase in the zebra finch and the two phases of listening and delayed production for swamp sparrow.

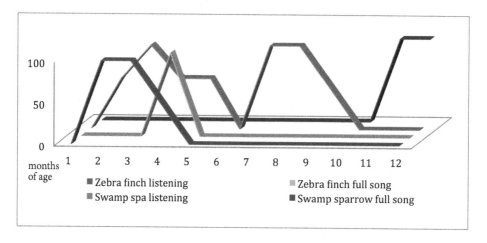

Figure 7.1 Example of sensitive phases and timing for full song for two types of songbirds, (after Hultsch & Todt 2004: 85)

As is shown in Figure 7.1, the zebra finch has a short sensitive phase span of between two and five months of age, and starts the sensorimotor production phase already at the age of five months.

The intra-specific differences for learning to sing are not only found in the age window of the juvenile, but also when it comes to demands on the tutor. Some species only learn if there is a live tutor, for example the father or a neighbour others are able to form the necessary auditory model on the basis of recordings. The zebra finch prefers a live tutor, but if there is none around, having some control of the song stimulus by pecking on a key, helps in acquiring songs from played back song (Adret 1993). If there is a choice between different tutors, the father is usually preferred, particularly when the father is housed with the mother. Otherwise, for example if another male is paired with the mother, he might be chosen instead of the biological father (Mann & Slater 1994, 1995). The white-crowned sparrows *(Zonotrichia leucophys)* are able to learn from tape in the beginning, but after around 50 days, they need a live tutor as a model in order to learn (Marler 1970a). The nightingale, on the other hand, needs a familiar tutor during the first period but learns from other in later periods. This behaviour is believed to help the nightingale to create the typical large repertoire of 200 or so songs – in order to achieve this one familiar tutor would not be enough.

Other examples are the chipping sparrows (*Spizella passerina*), who do not sing the song of the father, but prefer the song of the neighbour (Liu & Kroodsma 2006). To make the story even more complex, studies on the chestnut-sided warbler (*Dendroica pensylvanica*) have revealed that they have two types of songs, of which only one is dependent of a life tutor, the other can readily be learned from tapes. "Each strategy involves a balance between internal, physiological constraints and external, social constraints" (Byers & Kroodsma 1992: 808).

Some bird species have an extremely open program and are able to pick up all kinds of different sounds from the environment. Well-known examples are *Psittaciformes* (see Box 7.1) such as parrots, galahs, and budgerigars, but also starlings and other oscines are able to imitate various sounds, from speech to music, dog whistles and sawing (Marler 2004a). The Northern mockingbird (*Mimus polyglottos*) is called polyglottos because of its ability to "speak many tongues". The Latin name polyglottos means "many-tongued" and is used to describe human multilinguals.

Pepperberg (1998: 385) suggests that the talent to imitate others is a useful tool for temporarily joining or repelling other species. The kind of sounds that are imitated is constrained not by difficulty of sound characteristics such as fundamental frequency or rhythm, but rather by lack of motivation and learning environment. In experiments teaching human language to a Grey Parrott (*Psittacus erithacus*), Pepperberg (1998, 1999) has demonstrated that the success of learning depends on several factors. It is more likely to succeed if the sound makes a difference, i.e. it has some meaning attached to it, and if it is used by others in social interaction. Pepperberg uses a method with three-way interactions, where one human is teaching a second human and the bird is included in the interaction, to ensure a socially relevant situation.

Vocal learning of birds is often compared to language learning in humans. The developmental stages, auditory feedback, and ability to imitate are three features that humans and birds have in common. A fourth characteristic could be parallels in genetic expressions. As mentioned in Chapter 3.4, Specific Language Impairment (SLI) has been suggested to correlate with disruptions of the gene FOXP2 (a point mutation in the forkhead box). Haesler and colleagues (Haesler et al. 2004, Scharff & Haesler 2005) have found the same gene FOXP2 to be expressed in zebra finches during their sensitive phases, and also in canaries during their song plasticity period. Other birds that are found to have the FOXP2 expression controlling brain development are hummingbirds and parrots (Wada et al. 2004). Furthermore, analyses of the brain structures reveal other similarities between songbirds, parrots and hummingbirds on the one hand and humans on the other hand. All these species have developed specialized forebrain regions that are different from the brains of species that are not vocal learners (Jarvis 2007). This opens up a new scenario for comparing learnability and language evolution, suggesting that human language draws on properties that are present in a wide variety of species.

Most studies on learning have focussed on songs and less is known about the learning of calls. Boughman and Moss (2003) identify two types of vocal learning: learned acquisition and social modification. They claim that the learned acquisition is dependent of a sensitive age and involves learning new vocalizations, whereas social modification occurs independently of age and involves an increased similarity of vocalizations rather than new acquisitions. Typically, birds learn song by learned acquisition and calls by social modification. The distinction is based on empirical results showing that song learning has a sensitive age: birds that are deafened or raised in isolation are not able to acquire the song, whereas calls can be modified also in adults. Furthermore, the modification of calls is more an effect of convergence within a group than a juvenile taking after the tutor. Boughman and Moss discuss a study of a flock of chickadee birds (*Parus atricapillus*) that had had no prior contact, where the calls converged and reached increased similarity already after a week of contact. After four weeks there was a considerable convergence effect. No single bird served as model, but the convergence took place on mean features of the calls, and could be observed in frequency measures. Dialect variation has been observed in calls as well as in songs, and birds living near a dialect border produce some calls from each dialect, i.e. they have reached a level where they are practically "bilingual". The possibility to converge to a group later in life probably has strong beneficial effects for the recognition of group members and smooth functioning of groups (cf. speech accommodation Chapter 3.1.1).

Individual recognition is imperative in seabirds such as guillemots (*Uria aalge*), which hatch in tight colonies and crowd together shoulder to shoulder on sharp cliffs. The juvenile starts squeaking days before hatching and learns to recognize the voices of the parents while still in the egg. Later, when they have taken the "big jump" down to the ocean, the juveniles can rely on the calls to find the parents waiting in the water (Tschanz 1959). Individual calls may also be learned later in life.

7.4.2 Dialectal differences

As birds learn their song from parents and neighbours, this leads to geographic variation. Chaffinches (*Fringilla coelebs*), that live in the same area have songs that resemble each other (Thorpe 1958). The functions of these dialects are yet unclear. As with human dialects, it has been suggested that dialects facilitate the social life, that the dialect creates a group identity to the members. Another possible reason for the variation is environmental factors, either because of the vegetation or because of the body constitution. The transmission of sound is different in open landscapes and dense woods. Dialects are found in many songbirds. Well-known examples are the chaffinch, the white-crowned sparrow (*Zonotrichia leucophrys*), and the corn bunting (*Miliaria calandra*). A dialect area may include between 30 and 100 males, each singing two or three song types (Catchpole & Slater 1995). The dialectal variation is rarely described as exhibiting particular features, but rather as a variation in repertoire, so that some songs are shared by many birds, whereas other songs are only sung by individual birds. For example, in a study of 42 chaffinches, it was found that one of the songs was sung by almost half of the birds, whereas there were 21 songs that only were produced by individual birds (Slater et al. 1980). In another example, 8 out of 10 black-capped chickadees (*Poecile atricapillus*) produced three different songs, two of which had added introductory notes, whereas two birds only sung two songs, i.e. the original and one with the introductory note (Gammon & Baker 2004).

Although the occurrence of dialects is well-documented, the dialect boundaries, and acoustic characteristics analysed in great detail, our understanding of its ecological and evolutionary function is still scarce. There have been many suggestions as to why dialects emerge in birds, and Catchpole and Slater (1995) summarize three different perspectives. One possible function could be a kind of social adaptation, by which a male gains in importance by imitating the song of an older and more experienced male in the neighbourhood (Payne & Payne 1977, Nelson & Soha 2004). Another suggestion is that dialects evolve to match the habitat, for example that a particular feature such as a rapid trill should be more used in open areas, whereas slower trills are used in dense vegetation. The third suggestion is that variation in song has its parallel in genetic variation. So far, this has only been found for birds living in isolated areas (e.g. Baker & Cunningham 1985). Empirical investigations have yielded different results. The fact that females choose males on the basis of their song has lead to speculations about the role of dialect in mating. In some species, e.g. the corn bunting, females avoid mating with males of the same dialect, which may function as a means to enhance genetic variation. In other species, such as the great reed warbler, females instead prefer the same dialect and seem to use song cues to avoid mating with immigrant males (Hasselquist 1998).

The function of dialectal diversity is even more complex in birds that use their songs both to defend the territory and to attract females. Then the dialect may be beneficial in territorial defence, but disadvantageous in attracting females from other areas.

7.5 Visual communication

Visual communication in birds has not attracted the same interest as vocal communication, judging from the number of published books and journal articles. However, this does not imply that the visual channel is not used. Visual signals are abundant, from colours of the mouth gapes, to ornaments such as head plums, chest shields and elongated tails, and spectacular movements such as the bower elaboration of the bowerbirds, or the ritual dancing performed among grouse courting at the lek places.

It has been suggested that rich colours indicate that the animal is in good health since it shows an absence of parasites (Maynard Smith & Harper 2003). The same goes for the acrobatic and energetic movements involved in courting.

7.5.1 Talking with the tail

One feature that has taken many different extraordinary shapes during evolutionary history is the tail of the bird. One of the most well known examples is the peafowl (*Pavo cristatus*). As pointed out by Darwin, the size of the peacock tail has a high price in terms of mobility and also in energy and nutrients involved in growing the plumage, but the value lies in mating success. In an experiment where some eye-spots were removed, the result showed that number of eye-spots is indeed important for reproductive success (Petrie & Halliday 1994). However, this claim has been challenged by observations suggesting that the train may be only one of many significant features. The peacock signal is rather a multi-modal one, where behavioural features, such as shivering are as important as the visual impression (Takahashi et al. 2008).

The length of the tail is important in many species, and it is easy to manipulate its length. For example, the male long-tailed widowbirds (*Euplectes progne*) have tails of about 50 cm (the female tails are only 7 cm). In a study where the tails were shortened in some males and the feathers used to lengthen the tails of other males to about 75 cm, Andersson (1982) found female widowbirds to prefer males with these artificial long tails. Other studies manipulating tail length have come to the same result, for example in species such as the shaft-tailed whydah *Vidua regia* (Barnard 1990), Jackson's widowbird *Euplectes jacksoni* (Andersson 1992) and the swallow *Hirundo rustica* (Møller 1990). Most studies have dealt with polygamous species, where a long-tailed male may mate with many females, which gives reproductive success. The same result is attained in the case of the monogamous swallow. The effect of the longer tails is earlier mating, which is favourable, since it may allow for two broods instead of one during the season. The reason behind the long tail being so attractive, despite the costs (hindering the flight performance, increased risk of predation) can for example be that the males communicate good physical condition or that their sons will inherit their attractive long tails.

7.6 Chemical communication

There are very few studies investigating chemical communication in birds. Since birds have glands at the base of their tail, with secretes that are used in preening of the feathers, they may use some kind of olfactory signals, and the lack of studies may be because "we have been so attracted by bird song and visual display" (Wyatt 2003: 21). There are, however a few studies on the olfaction in birds, demonstrating for example that mallard ducks (*Anas platyrhynchos*) use chemical communication for reproduction. In a study where the olfactory nerves were cut, the sexual behaviour of the duck was reduced (Jacob et al. 1979). Tube-nosed birds like albatrosses (*Diomeda exulans*) and Wilson's storm-petrels (*Oceanites oceanicus*) are famous for their sharp sense of smell and they are able to locate food from long distances even in darkness (Hutchison & Wenzel 1980, Nevitt 2000). It may be assumed that these birds use the olfactory capacities for communicative purposes, but studies remain to be done.

7.7 Birds and humans

7.7.1 Teaching language to birds

Considering the similarities in early vocal learning, it could be assumed that each species learn the vocalizations of the others', but this is rarely done. However, imitation of human language is known from many birds with open-ended learning, particularly parrots (*Psittaciformes*). They are able to learn new vocalizations throughout life and can imitate all kinds of sounds, such as train whistles, musical instruments, and human speech. An ability to imitate others has been assumed to be advantageous in the kind of large mobile flocks that are typical to parrots. Young birds forming their first flock quickly learn a shared call that functions as a marker of group identity, and they are then able to single out the voices of flock-mates from the sounds of other voices in noisy environment (Farabaugh & Dooling 1996, Pepperberg 2004).

Their remarkable ability to imitate is probably the reason why parrots are so successful when it comes to learning human language. In studies of the African grey parrot (*Psittacus erithacus*), Irene Pepperberg (1998, 1999, 2004) has shown that parrots are capable, not only of imitating sounds, but also of comprehending and using language in an appropriate way. For example, the grey parrot Alex had a vocabulary of 50 objects, 7 colours, 5 shapes, and 6 numbers. He could use language referentially and answer questions about shapes, sizes, numbers and colours of objects. Furthermore, he understood language functions, such as the difference between labelling and requesting objects, and was able to alternate between different communicative roles such as asking questions and giving answers. Another interesting finding is that Alex engaged in monologic speech. Monologues are commonly found in the early language development of human children, often as sound play, and this was found to occur in several of the parrots in Pepperberg' studies. The parrot Kyaaro, also trained by Pepperberg, gives an extraordinary example (see below), imitating a whole dialogue, using the different voices of the trainers.

Interestingly, Kyaaro, another grey in training who was less accomplished than Alex, used entire dialogues in solitude, reproducing two different trainers' voices and synthesizers sounds, as well as using his own voice; a typical example would be as follows:

- "Listen, Kyo" (in the voice of trainer 1).
- "Click, click, click, click (replicating the synthesizer).
- "How many?" (in the voice of trainer 2).
- "Four" (in his own voice).
- "Good boy!" (trainer 1 or 2). (Pepperberg 2004: 367)

The method Pepperberg uses is based on the assumption that social interaction is the optimal context for language learning, in parrots as well as in humans. The model is labelled the M/R (Model/Rival) model, since it, apart from the trainer, also involves another person that at the same time acts as model for the response and a rival for the trainer's attention. Thus, the setting involves three participants, two humans and a parrot. The humans communicate about objects, qualifiers, quantifiers and occasionally on actions, and the parrot observes the others, as well as taking his turn. Below an example of a training session is presented.

Participants:
A: Alex; I: Irene Pepperberg; K: Kimberley Goodrich, another trainer

- I: (Acting as trainer): Kim, what color? (Holds up green triangular piece of wood)
- K (Acting as model/rival): Green three-corner wood.
- I: (Briefly removes object from sight, turns body slightly away.) No! Listen! I just want to know *color*! (Faces back toward K. Re-presents object.) What *color*?
- K: Green wood.
- I: (Hands over wood): That's right, the color is *green*; green wood.
- K: OK, Alex, now you tell me, what shape?

A: No.
K: OK, Irene, you tell me what shape.
I: Three-corner wood.
K: That's right, you listened! The shape is three-corner; it's a *three-corner* wood (Hands over wood.)
I: Alex, here's your chance. What color?
A: Wood.
I: That's right, wood; what *color* wood.
A: Green wood.
I: Good parrot! Here you go (hands over wood). The color is *green*.
(Pepperberg 1999: 57)

The method is clear from this example, Alex takes part in an interaction, where he is expected to contribute with his parts. The human speakers interchange the roles of trainer and model.

As natural as the interaction may seem, it is not really what is found in conversations with human infants (see Chapter 3.2). Rather, it may be compared to what takes place in second and foreign language classrooms. Pepperberg & Schinke-Llano (1991) suggest that both humans learning their second language, and parrots learning a human language can be considered cases of exceptional learning. In this kind of learning, the input and interaction must be both qualitatively and quantitatively superior to what a young child encounters, when learning a first language.

Another tradition in the teaching of human language to birds is to use syntax. One of the hallmarks for human language has been assumed to be capacity for syntactic recursion (see Chapter 3.3.5) and there are many attempts to teach this feature to nonhumans, both to birds and to other animals. One of the more successful is a study on starlings (*Sturnus vulgaris*) where syntactic abilities are reported (Gentner et al. 2006). The starling is a good candidate for such a teaching experiment, since it is one of the songbirds, mentioned above, that has an open-ended learning program. Gentner et al. (2006) trained starlings to recognize recursion, that is embedded strings of the kind *This is the malt that lay in the house that Jack built* (see Chapter 3). The starlings were exposed to stimuli with 'rattle' and 'warble' motifs in different patterns, for example 'rattle, rattle, warble, warble' that is AABB (A^2B^2), and 'rattle, warble, rattle, warble' that is ABAB $(AB)^2$. Out of the 11 starlings that were trained, 9 managed to differentiate between the patterns. Moreover, the birds did not only differentiate between the stimuli they got but they were also able to generalize to new stimuli, demonstrating that they recognized the patterning rules behind each sequence. It has been claimed (e.g. Marcus 2006) that recursion as a computational tool is not a feature of human language, but goes beyond the human language faculty, and may be part of more general capacity found also in other species.

7.8 Summary

This chapter has presented selected findings from studies in bird communication. In the first section, the functions of calls and songs were discussed, with some aspects, such as referential communication, described in more detail. Then, communication in birds was discussed from the perspective of ontogenetic development. We considered some aspects of song development, taking aspects such as sensory phases, sensori-motor phases, developmental stages and the need for live tutor into account. The section on songs provided an account of interactive singing, in counter-singing and duets. Finally, results on the teaching of human language to birds were presented.

7.9 Suggested readings

Kroodsma, D. E. & Miller, E. H. (Eds). (1996). Ecology and Evolution of Acoustic Communication in Birds. Ithaca, N. Y.: Cornell University press.
This book gives a thorough account of acoustic communication in birds. Issues that are discussed in the individual chapters include: how genetic and environmental factors interact in the development and production of song, how birds communicate in a noisy worlds, and how playback is used as a method to investigate meanings of vocalizations.

Marler, P. & Slabbekoorn, H. (Eds). (2004). Nature's music: the science of birdsong. Amsterdam: Elsevier.
Marler and Slabbekoorn have put together a masterpiece in their edited book Nature's music. In the introductory chapter Peter Marler gives a personal account of 50 years of the science of birdsong, of which he is one of the leading experts, in his chapter "Science and birdsong: the good old days". This chapter is followed by chapters on functions of singing, sensitive periods, plasticity and other issues in song learning, bird calls, brains and birdsong, birdsong and music, evolution of birdsong, teaching language to birds, and conservation. The book comes with two audio CDs with recordings donated by different researchers (for example the original recordings by W. H. Torpe) and colour illustrations.

Pepperberg, I. M. (1999). The Alex studies. Cognitive and Communicative Abilities of Grey Parrots. Cambridge. MA: Harvard University Press.
This book gives a panoramic view of Pepperberg's groundbreaking work with the Grey parrot Alex over a period of twenty years. The methodology and results are described in great detail. The findings of Alex's cognitive and communicative abilities are placed in the context of other research in biology, comparative and cognitive psychology, linguistics and neurobiology.

Chapter 8

Discussion and outlook – why language?

8.1　Introduction

In the previous chapters we have presented studies of human and animal communicative behaviour. We have examined communicative functions, forms and communicative development in the young. What conclusions can be drawn from this exercise? In this chapter we will give a synthesis of different issues. First some general aspects of comparing humans to animals will be discussed.

As is often pointed out, the human-animal comparisons are complicated by the fact that when we deal with communication in humans and animals we interpret everything through our own language. We analyze social behaviour through our own cognitive language-connected filter – and this filter is also present when we categorize animal communication. "Perhaps we are predisposed to see other species' communications through the filter of language metaphors because language is too much a natural part of our everyday cognitive apparatus to let us easily gain an outside perspective on it". (Deacon 1997:34). The American linguistic Bickerton even goes so far as to claim that our cognitive ability, for example to solve the problem about communication, has developed from our linguistic ability, and this makes it even harder to get round our own language in trying to understand that of others' (Bickerton 1995:160). The comparisons between humans and other animals usually start with the human as the baseline, and then other animals are compared to that baseline. There have been attempts to take the other perspective and observe human behaviour from the perspective of other animal behaviour (e.g. Eibl-Eibesfeldt 1972, 1986), but that is rather an exception to the general tendency. This bias may be problematic, since we may miss out on important communicative functions and features of other systems. Furthermore, as Millikan (2004) points out, there is a risk that we miss something important in characterizing animal communication from the point of view of human language. If animals do not represent certain contents, or employ particular behaviours, this may be because they have no use for them. Another problem with all comparisons between humans and animals is that it is often taken for granted that, if we find similarities, they function in the same way. On the contrary, it is quite possible that something that looks similar has different functions in two biological species. "This means that from studying monkey brains we can't infer for sure how the human brains perform even on the "low" level of the way classes of neurons function. This is definitely not a good news, as the majority of the neurological studies of monkeys and primates are made with an eye that the human brain performs the same way" (Stamenov 2002: 269). Another problem is the restrictions on the naturalness of the communicative situation in studies of animal behaviour. DeVore (1973) said that the early studies of captive nonhuman primates should be compared to human behaviour "in a maximum-security prison", and Kroodsma pointed out that what a bird is doing in a laboratory only shows what a bird is capable of doing in a strange environment (Kroodsma 1996: 4, cf. above).

Since most comparisons have had human language as point of departure, investigations have been carried out on features that are regarded as typical to human language, for example referential functions (alarm calls, food calls), shared attention (pointing, eye contact, gaze following), alignment (accommodation or synchronization of behaviour), syntactic patterns (elements occurring in predictable orders), dialects (different forms in different groups), and learning (critical periods). However, despite often common targets, there are striking differences in research questions and methodology. For example, playback, which is a favourite method in research of animal communication, is rarely, if ever, found in investigations on human communication. Studies on how meanings are shaped by the context (Chapter 3.3.1) are totally absent in animal communication research, and longitudinal studies of emerging communicative skills which are often used in research on human communication, are rare in studies on animals.

8.1.1 Social life and communication in humans and animals

Despite the richness and large variation in communicative channels, the differences in how behaviour is perceived (hearing range, eye sight and other sensory recepients) there are some general similarities between communicative behaviour in different species. The social group determines the need for communication – as was noted already by Darwin (1872). According to the 'social complexity hypothesis' (Freeberg et al. 2012) there is a correlation between complex social life and complex communication. Furthermore, egalitarian groups have more complex communication than hierachical groups, since individuals in egalitarian groups have more interaction and confrontations. In an egalitarian social group of the fission-fusion kind, there is a need to express group identity as well as individuality. This calls for fine-grained expressions of group solidarity, coherence and cooperation. Group coherence can be easy to observe in the synchronization of for example postures and movements (Meltzoff 1981, Dimberg et al. 2000, Rotondo & Boker 2002, Bråten 2002).

Many of the behaviours that we have presented from different species show traces of synchronization. One example is the antiphonal calling, with alternating turns, found among elephants (Soltis et al. 2005) and humans (laughter; Smoski &

Bachorowski 2003). Another example is when the participants vocalize together, for example human singing, duets in gibbons (Geissmann 2003), vocal sharing in Campbell's monkeys (Lemasson & Hauberger 2004), communal howling in wolves (Harrington & Asa 2003), and vocal chorus in group-living birds (Radford 2005). These are examples of acoustic nonverbal correlates to synchronization of visual and chemical communication and alignment in dialogues.

A convincing example of the robust interactive basis of language use comes from a study on the neural mechanisms underlying speech perception, where the tongue movements of listeners were measured. The results demonstrated that there is an activation of tongue muscles also during passive listening. This was particularly strong when listeners heard sound that required strong movements, for example a double 'r'. (Fadiga et al. 2002). Similarly, studies on mirror neurons have shown that, when someone observes an action, an internal copy in the brain of that action emerges (di Pellegrino et al. 1992, Rizzolatti et al. 2002). This facilitates understanding of others' actions, and communicative behaviour. It may also explain contagious synchronization.

8.2 What is so special about language? Revisiting Hockett's predictions

When Hockett (1968:18) claimed that "No species except our own has a language", a generalization that has become a classic, he added the remark "This may be disproved at any time by new zoological discoveries". One may wonder – has it been disproved by new discoveries now? The following sixteen features were suggested to define language, and only if all these were present, the communicative system may be called a language. The features are (1) using the vocal-auditory channel; (2) broadcast transmission with directional reception; (3) rapid fading; (4) interchangeable; (5) complete feedback; (6) specialization; (7) semanticity; (8) arbitrariness; (9) discreteness; (10) displacement; (11) openness; (12) tradition; (13) duality of patterning; (14) prevarication; (15) reflexiveness; and (16) learnability (Hockett 1968: 8–13).

Of these design features, some have been challenged in human languages (number 10, displacement; Everett 2005), and many others have been found to occur also in other species, to a larger or smaller degree. Some features are hard to investigate objectively, for example openness (the potential to talk about anything), specialization (that the system is well designed), prevarication (that is is possible to tell lies) and reflexiveness (that the system allows communicating about the system). However, the other features are found in one or several species. Vocal-auditory communication (1) is common in many species, broadcast with directional reception (2) is found for example in chicken (Evans & Marler 1994), and rapid fading (3) is found in all vocal communication. Interchangeability (4) is found in gibbons (Geissmann 1983), and complete feedback (5) is found in songbirds (Marler 1970a, b). Semanticity (7) and arbitrariness (8) are found for example in alarm calls of vervet monkeys (Struhsaker 1967) and domestic chicken (Marler 2004b). Discreteness (9) is found in starlings (Gentner et al. 2006). Displacement (10) is found in bees (von Frisch 1954). Tradition (12) is found in birds (Thorpe 1958), and duality (13) is possibly also found in birds (Gentner et al. 2006). Learnability (16) has been found in many different species, such as nonhuman primates, dogs and birds. However, as far as we know it is only in human language that all features have been reported to be present.

Hockett is not the only one to suggest discriminating features. Using language development in children as a starting point, Oller (2004) proposes a hierarchy of features, with the semantic/illocutionary decoupling as a key factor. To be able to separate the semantic meaning (e.g. the comment "There is a lion.") from the illocutionary force ("There is a lion!") requires something else than just be able to refer to some external object.

> The development of a lasting model to replace the Hockett scheme will require a clear theoretical formulation of the steps that yield the human vocal progress in the first months of life, as well as comparative empirical demonstrations of vocal communication capabilities in both human infants and nonhuman primates. Of course the model will need to account for many other species besides the primates, and it seems that in some cases, at least other species (parrots and song birds provide good examples) may surpass the nonhuman primates in vocal command including, perhaps, in the ability to decouple "words" from particular illocutionary forces, after "language" training.
> (Oller & Griebel 2004: 329)

Looking into the illocutionary force of utterances implies a shift in perspective from looking at communicative functions and structures in isolation to a socially grounded view on communication.

The referential function, which is assumed to be so typical to human speech (based on joint attention and conventionalized, arbitrary form-meaning relationship) has often been investigated in isolation from the social context in animals. By recording vocalizations and playing them back to a group of animals, it has been possible to study correlations between sound structure and reactions. Referential calls are found to be part of the communicative repertoire of many different species, from prairie dogs (Slobodchikoff et al. 1991) and red squirrels (Greene & Meagher 1998) to domestic chicken (Evans & Marler 1995, Forkman 2000). The calls are assumed to have fixed meanings, and when the same vocalization can mean different things to different listeners (for example a bird's song can attract a female and repel a male) this is mentioned but not taken up for serious discussion.

8.3 Why and how did language evolve?

As mentioned in the introductory chapter, there has been an outburst of books and journals dealing with the evolution of human language. One issue that has emerged in the discussions about the origin of language is if it is possible to find correlates to language, other phenomena that distinguishes humans from other species. The questions that are asked are not only in what way human language is different from other communicative systems, but also why and when language evolved at all. Are there any parallel changes in human physiology? What about the role of cooperative breeding?

8.3.1 Why and when did it happen?

In Johansson (2005) five hypotheses about why language came about are discussed. Firstly, our ancestors needed language to coordinate joint hunting. This hypothesis builds on the assumption that herbivores can find food by themselves, whereas meat eaters have to coordinate hunting. They have to find the prey, kill it, and share the meat. It is easy to understand that language is useful both at the planning stage, which is a future-directed co-operation, and afterwards, when persuading the others to share the gain (Brinck & Gärdenfors 2003). However, language does not seem to be a necessary prerequisite, since also other group-living species, such as wolves, lions and nonhuman primates succeed in coordination of hunting.

A second hypothesis is that language evolved together with tool-making. Tool-making is associated with human culture, planning ahead and teaching traditions to next generation. However, also other animals make tools (e.g. chimpanzees, bonobos). Even more species use tools for various purposes.

A third proposal is that language evolved for sexual selection. This assumption is based on general principles of evolutionary success. The peacock has its tail, and humans talk in order to advertise themselves (Darwin 1872/1965, Burling 2005). An interesting difference between humans and other primates is that the human female has a concealed ovulation, and monthly menstruations. Since ovulation is not apparent and since a fertile female menstruates, the male has to be more attentive to the signals (Power 1998).

The fourth hypothesis has to do with the fact that human children are more dependent at birth than infants of many other species, and they also take longer time to reach adulthood. Features that emerge early and before the emergence of language, such as joint attention, gaze following, and imitation, indicate that human infants are open to learning from the adult interactants. It seems plausible that language could play an important role in teaching. However, nonhuman primates also demonstrate joint attention etc., and they manage without language.

The fifth hypothesis is related to the fourth, in that it has to do with the complexities of social life. Language may be a product of the demands of social life, rather than as a practical means to find food. Several authors have suggested that when the human groups grew larger, language, in form of gossip, replaced allogrooming (e.g. Dunbar 1996). Box 8.1 summarizes some pros and cons of different hypothesis.

Since there are no visible traces of language evolution, we have to look around for other related phenomena, for example tool-making, brain size, and cave painting, in order to date the evolution of language. Regular tool-making came about around 2 million years ago, and the cranial capacity increased between 2 and 1.5 million years ago. It seems reasonable to assume that some kind of language came about around that time. Another major change in the human brain occurred some time during the past 400.00 to 200,000 years when the upper limit of the cranial capacity increased

Box 8.1 Hypotheses as to why language evolved (from Johansson 2005)

FACTOR	PRO	CON
1. Group hunting	+ Economically important + New ecological niche for *Homo*	− Other hunt in group as well, primates as well as wolves, lions, etc.
2. Tool making	+ Major factor in human evolution + A structured sequential activity – much like language	− Other primates make tools as well − Much tool making is a solitary activity, what is the use of language?
3. Sexual selection	+ Language use important for mating success + Human mating more complex, e.g. concealed ovulation	− Other primates have mating success too − Mating system of chimps also complex (but not concealed ovulation)
4. Child-care, and teaching	+ Humans have extended childhood + Language useful for teaching + Teaching rare in other primates	− Other primates don't teach, the need for teaching must have evolved before language
5. Social relations	+ Social relations is the primary topic in language use + Complex network of friendships, alliances + Human social groups are very large + Social communication valuable in social relations	− Other primates have complex group networks too

again (Davidson 1999). However, a large brain is not necessarily the indication of language. It is costly to maintain and it has been suggested that the Neanderthals actually had a larger brain than the *Homo sapiens*.

The study of language evolution is an interdisciplinary area, involving research questions and methodologies from biology, genetics, archeology, anthropology, psychology, linguistics and, during the last decade, also computer modelling. By computer simulations it is possible to test out hypotheses, replicate experiments and manipulate the conditions so that different components are accessible for analysis. In this way, simulating language evolution can sort out some of the complexity involved. Parisi and Cangelosi (2002) describe a model where both language development, evolution and change are accounted for. In this simulation, organisms are interacting with both environment and conspecifics, and it is shown how the emerging language influences the categorizations and actions of the organisms. Experiments like this, and also experiments with robots, open up for new ways to search for the answer to the questions of why and when language evolved.

8.3.2 Are there any parallel changes in human physiology?

There are several anatomical developments that may have had an impact of language evolution. For example, bipedalism, changes in the hand and facial muscles, growth of the brain, and lowering of the larynx, are changes that have been suggested to be adaptations that made language possible, or that developed in parallel with speech.

Relative to the body size, humans have a large brain. This is assumed to have opened up for possibilities to regulate syntactic patterns. Areas that are often mentioned as being involved in speech are Broca's area and Wernicke's area (named after the neurologists Paul Broca (1824–1880) and Carl Wernicke (1848–1905) who discovered correlations between brain injuries and language problems). However, the assignment of certain functions to specific areas has caused a lot of controversies. Evidence from studies using techniques such as PET-scanning (positron emission tomography) and fMRI (functional magnetic resonance imagery) demonstrate that many cortical areas operate simultaneously, with specific locations for elementary functions. This is possible through interconnections between cortical systems. As the neurolinguist Sydney Lamb writes, it is through a growing net of connections that language is acquired.

> And so if we ask what is innate (…), maybe the answer is that the most distinct innate features of our cortices, those which makes us most different from other mammals, are the increased abundance of cortical columns and of their interconnections, including the fantastic possibilities of interconnection made possible by the long-distance axon bundles, and the really wonderful fact that only a small minority of them are innately hard-wired. Most of our great cortical power comes precisely from the fact that the

> great preponderance of information processing structure is not innate. We are therefore able to learn to do things, to recognize things, to ponder about things, to enjoy things, than could never have been foreseen. (Lamb 1999:372)

Another suggested adaptive development is the change of the human larynx. As mentioned in Chapter 3.2.3, the larynx undergoes a dramatic development in early infancy, when it descends and allows the tongue to move in a way to produce typical speech sounds. A second change occurs at puberty, when there is another descent in the larynx of males. This development in ontogeny has been proposed to have a direct parallel in phylogeny, and to be one of the adaptations that once made speech possible (Studdert-Kennedy 1998). However, speech is not a necessary prerequisite for language, since sign language can be used to express all ideas that are expressed through speech, so the low position of the larynx is not what is needed for language. Furthermore, Fitch (2002) suggests that the descent of the larynx may have other functions than speech. A postpubertal change of the position of the larynx is present also in other animals, for example the male deer (red deer *Cervus elaphus*, and fallow deer *Dama dama*). This permits them to produce low-pitched vocalizations to threaten rivals during the breeding season. In other words, they are using the frequency code suggested by Morton (1977), and exaggerating their size. The same function may have been what triggered the descent of the larynx in the early hominids, and only later the increased size of the vocal tract was exploited to produce speech sounds.

There are also other changes in the human physiology that may have a direct effect on the evolution of language. As mentioned earlier (Chapter 4.4) the region around the eyes is an area where humans differ from other primates (Emery 2000). For example, humans have large foreheads, cheek fat and eyebrow hair, whereas nonhuman primates have low wrinkled foreheads and no cheek fat. The human eyebrows provide a frame to the eyes, and the eyeball is more flexible and has a well-developed sclera surrounding the iris. This design may facilitate shared attention, since it is easy to detect where people are looking. Shared attention in turn, is a prerequisite for referential communication.

A feature that has been suspected to be an adaptation to speech is categorical perception, i.e. that speech sounds are perceived as distinct entities. However, this property is not unique to humans. Results from studies of chinchilla (Kuhl & Miller 1975, rhesus macaques (Kuhl & Padden 1983), dogs (Fukuzawa et al. 2005), and birds (Dooling et al. 1995) demonstrate that other species are able to differentiate between human speech sounds (for example between *la* and *ra*). This means that the mechanism underlying categorical perception is part of an auditory system shared across different animal species. Furthermore, there is evidence that other species can even detect the difference between human languages. In an experiment where human newborns and cotton-top tamarin monkeys were exposed to speech sounds from two different languages, both the human infants and the monkeys showed an ability to discriminate between languages (Ramus et al. 2000).

8.3.3 Is language a result of cooperative breeding?

Cooperative breeding – where alloparents assist in providing and caring for the young – is known from both mammals and birds. It occurs in about half of the non-human primates. However, in our closest relatives, the Great Apes it is the mother that takes the resposibility for the offspring for several years. This means that humans are exceptions to Great Ape behaviour in having alloparents. In her book *Mothers and Others*, the anthropologist Sarah Blaffer Hrdy describes the situation as follows:

> Women are just as prone as other apes to worry about the well-being of new babies. But what hunter-gatherer mothers do not do postpartum is refuse to let anyone else come near or hold the baby. This is an important difference. A brief survey of caretaking practices across traditional hunting and gathering people – the closest proxies for Pleistocence hominins we have – reveals that even though nomadic foragers differ in where and how they make a living, babies are universally treated with warm indulgence. Hunter-gatherers are no different from apes in this respect. Babies are never left alone and are constantly held by someone, but that someone is not invariably the mother. Human mothers are just as hypervigilant; they are just not so hyperpossessive. From the outset a human mother will allow group members (typically relatives) to take and hold the baby. (Hrdy 2009:73)

It can be speculated whether the cooperative breeding system, with the need to be taken care of and communicate with different individuals is what drove the human communicative system to develop into a symbolic language.

8.4 Learning intraspecific communication – not only for humans

It has often been assumed that intraspecific communication is inborn in animals but learned in humans. However, when looking closer to what happens during early development, we find that for example songbirds, some nonhuman primates and humans display many similarities in the development of communication. Juvenile birds and human infants have an early sensory period when they perceive and store the sounds and syllables that will be used later, and then a sensorimotor period when they practise. The sensory period is a phase of remarkable openness to perceive and memorize vocal models. This period has been suggested to be limited in time, both for birds (Thorpe 1958) and humans (Lenneberg 1967). For songbirds, the length of the sensitive period differs between species (and there are also open-ended learners), but it generally takes place during the first six months. For human infants there have been different proposals. It is not possible to undertake the type of experiments that are made with birds, so data either comes from case studies of socially isolated children, studies of children with brain injuries, or from studies of bilingual language learners. Humans retain the capacity to learn language over the life-time, but there are differences in learning the first and learning the second language (in onset and final state). If the language is acquired before a certain age, it is acquired as a first language and will lead to native competence, otherwise it is a second language. From a macroperspective, the developmental process can be seen as a combination of input, memory and a self organization system (Davis 2011). This goes for human language as well as animal communication. During the sensorimotor period songbirds, marmosets and human infants produce sounds and match them to a memorized target, through an auditory loop. Clear developmental stages have been identified for birds as well as for nonhuman primates and humans. Songbirds start with subsong, then plastic song and finally crystallization (se above). The young marmosets start with a subset of the sounds in the adult calls, which gradually become more adult-like. The early productions function as vocal practice, and infants that babble more frequently have better formed vocalizations earlier (Snowdon & de la Torre 2002). Young children start with babbling, then one-word utterances and after that, multi-words utterances. Second language learners do not start with babbling, but in other respects the developmental stages bear many similarities to first language learning (see Chapter 3 for more details). What is alike is that young birds, nonhuman primates and humans practise by first using all possible sounds, then restrict the sound inventory to what is used in the environment and then follow fixed developmental stages towards the target.

One difference between humans and the other species is the eye gaze behaviour. In early human parent-offspring interaction (as in second language acquisition), mutual gaze is the most prominent feature, whereas this is not reported to be as striking in other species. This makes sense, since joint attention is assumed to be what leads to the form-function mapping and symbolic word meanings.

8.4.1 Child language and animal communication

As described in Chapter 3, children go through well defined stages of phonological, lexical and grammatical development in their course towards adult language proficiency. Interestingly, when researchers set out to compare characteristics of human language to animal capacities, they often pick the earliest features in child development, and not the later. An overview of studies reveals that the aspects that are investigated in animals are the ones that emerge in children before the age of two years. Box 8.2 shows aspects of language acquisition that have been investigated for children of different ages, and compared to skills in other species.

Box 8.2 Areas of interest in human communication of different ages

AGE SPAN	FOCUS OF INTEREST	RESULTS	REF	INVESTIGATED IN OTHER SPECIES
0–2 years	Input	High pitch and exaggerated vowels are typical in speech to infants	1	Yes (humans to dogs, cats) (Burnham et al. 2002)
1–4 months	Categorical perception	Infants discriminate between languages	2	Yes, cotton-top tamarin monkeys (Ramus et al. 2000)
6–8 months	Phonemic contrasts	Japanese children perceive contrasts at 6–8 months bur not 10–12 months	3	Yes, rodents (Kuhl & Miller 1975), dogs (Fukuzawa et al. 2005), birds (Dooling et al. 1995)
3–8 months	Anatomy	Larynx lowering around 3 months, before infants more like chimpanzees than adult humans	4	Yes, chimpanzees (Nishimura et al. 2003)
0–1 years	Babbling	Universal stages found	5	Yes, birds (Podos et al. 1995), pygmy marmosets (Elowson et al. 1998)
15 months	Word meaning	Fast mapping of relation between form an content	6	Yes, dogs (Kaminski et al. 2004)
18 months	Grammar emerge	Grammar starts with items, slots for new words	7	No
1;6–4;0 years	Grammar	Semantic relations, developmental stages based on MLU	8	Yes, language-taught chimpanzee (Gardner & Gardner 1969)
2–4 years	Word meaning	Overextending meanings	9	No
2–4 years	Words and grammar	Language-specific structures, prepositions	10	No
3 years	Subordination	Comprehension and production of relative clauses	11	No
4–7 years	Word inflections	Production of plural endings to unknown words	12	No
5 years	Words and grammar	Language-specific codings of events	13	No

References: 1. Kuhl 1999   2. Eimas et al. 1971   3. Tsushima et al. 1994   4. Kent & Miolo 1995   5. Roug et al. 1989, Vihman et al. 1986, Boysson-Baridies et al. 1989   6. Schafer & Plunkett 1998   7. Tomasello 1992, 2003   8. Brown 1973   9. Clark 1978   10. Bowerman 1978, 1994   11. Håkansson & Hansson 2000   12. Berko 1958   13. Slobin 1996.

8.5 Can language be taught to nonhumans?

This brings us to the last issue, the teaching of language to other animals. Language is acquired by human children during their first three years of life, together with their socialization into the group that speaks that language (Halliday 1975, Gaskins 2006, Ochs & Schiffelin 2001). They do not need teaching to learn it. When people learn second languages, however, they often do it by help from teachers and a school curriculum. For practical reasons, this method is quite common across the world, but it is not necessary for language learning to take place. Language is not only as a syntactic-logical system, but also an act of cooperation, where the participants align to each other and create meaning in the context, and this can be accomplished also without explicit teaching. In studies of language teaching to nonhuman primates this view is rarely expressed, and in most cases there is no meaningful context and to social group to interact with, and accommodate to. The chimpanzees seldom observe others use sign or plastic bricks in social interaction. Generally, the animals in teaching experiments have been exposed to teaching methodologies based on deterministic stimulus-response patterns, and they are not given the possibility to react to a natural social environment, which forms the basis for human children learning language. The lack of meaningful interaction may explain why many of the experiments with traditional teaching methods have failed, whereas experiments using an interactive approach (e.g. Savage-Rumbaugh & Lewin 1994, Pepperberg 1999, see also Lyn 2012) have been more successful.

The success of nonhuman primates that are taught language is usually compared against measures from human children acquiring their mother tongue, despite the fact that, strictly speaking, the human language is a second language to a nonhuman primate. This concern turns up in Segerdahl et al. (2005) when the authors make an attempt to disentangle 'a language' from 'a specific language'. The first step for Kanzi was to learn that there is a linguistic system (different from his own), before he could enter into a specific language, English, which then became his second language. It is also possible to claim that is was his third or fourth language, taking the bonobo communicative system as the first (he was six months when he was first exposed to language training), the linguistic system as second, and keyboard symbols and spoken English as third or fourth, respectively. According to Joseph (2004), first language users use language both for identity, thinking and for communicating, whereas second language users use language only for communication. If the aim of teaching language to nonhumans is to teach them how to communicate, then their progress should be compared to human second language learners, and not to children acquiring their first language.

In an article from 1979, Terrace et al. asked the question "Can an ape create a sentence?", and answered the question by "No". But if we instead ask: "Can an ape express an emotion?" there is a different answer. Noone who reads the passage in Terrace

(1979), where it is described how he visits Nim after a year of separation, can doubt that Nim is able to express feelings, in fact both verbally (by signing), and nonverbally (by hugging).

> As soon as I turned to greet him and he realized who I was, Nim's mood changed instantly and dramatically. He let out a wonderful shriek, leaped into my arms, signed *hug*, and gave me a tremendous chimp embrace. It was wonderfully familiar. If I had any doubts about his remembering me, or whether his reaction would be cool, they were quickly dispelled. A few hugs later Nim pulled away and began to groom me. When I put him down for a moment he looked a bit puzzled and troubled. After shaking his hands back and forth a few times he again signed *hug* and came back into my arms.
> (Terrace 1979: 229)

8.6 Summary

This chapter has summarized some of the aspects that are touched upon in the previous chapters, for example how communication is shaped by the social group, how social identity is expressed in synchronization, and how meanings are created in the context. Issues in the evolution of language have been presented. Hockett's design features have been revisited, with the conclusion that these features, suggested as characteristic for human language, are not uniquely human, but to larger or smaller degrees also present in animals. However, only human language holds all the features simultaneously. The cooperative breeding system has been presented as one of the possible factors behind the evolution of human language. Finally, methods and results from the teaching of language to other animals have been discussed. We find that only humans seem to be interested in learning and using language. The strong (unstoppable) motivation in young infants to learn language and to share the social life with other humans, has not been observed in other animals. Despite teaching, they do not seem to be able to reach beyond the performace of a two-year old child.

Much research is needed before we can understand this, as well as many other mysteries of the communication in humans and other animals.

8.7 Suggested readings

Enfield, N. J. & Levinson, S. C. (Eds.). (2006). Roots of Human Sociality. Culture, Cognition and Interaction. Oxford: Berg.
This book presents a multidisciplinary perspective of social interaction, drawing on research from ethnology, sociology, anthropology, linguistics and biology. The book consists of five parts; human interaction, psychological foundations, culture, cognition, and phylogenetic perspectives, all combined to form a coherent framework of human sociality.

Cangelosi, A. & Parisi, D. (Eds.). (2002). Simulating the evolution of language. London. Springer-Verlag.
Computer simulations makes it possible to test hypotheses on language evolution, development and change under controlled conditions. This book presents simulation work about the evolution of signalling systems, the development of dialects and the ontogeny and phylogeny of language.

Johansson, S. 2005. Origins of language. Constraints on hypothesis. Amsterdam. Benjamins.
In Origins of language different hypotheses are presented and evaluated from the perspective of how consistent they are with the available data. The book is written in an engaging and accessible style and gives the readers a clear overview of existing hypotheses of the evoution of language.

Kappeler, P. M. & Silk, J. B. (Eds.). (2010). Mind the gap: Tracing the Origins of Human Universals. Berlin. Heidelberg. Springer-Verlag.
This edited book discusses the "nature and width of the gap that separates humans from primates and other animals" (p. 3). The question is dealt with in the book by primatologists, anthropologists, biologists and psychologists. The "gap" is suggested to be behavioural, and studies of behaviour are in the centre of the book. The chapters deal with for example intergroup relations, conflict and dominance behaviour, communication in human and non-human primates, the basis for human culture, and cooperative breeding.

Glossary

Accommodation – adjustments that people make to their speech and nonverbal behaviour, influenced by how others behave

Adaptor – One of the categories for the analysis of nonverbal behaviour in the model suggested by Ekman & Friesen (1969). An unintentional gesture, for example scratching the nose, functioning as outlet of nervousness

Alignment – the way speakers adapt to each other during a conversation. For example when someone reuses the words or structures of the previous speaker

Allogrooming – also called social grooming. Nonhuman primates are often seen cleaning the fur and removing tics from another individual

Altricial – Altricial young are born helpless and require a lot of attention. (Latin root *alere* = to nurse, to rear to nourish)

Analogous comparison – between nonrelated species that share some behaviours. Opposite to homologous

Audience design – when the form of a vocalization differs according to which individuals are present

Aphasia – inability to use or understand language due to a brain damage

Auditory communication – communication that relies on hearing

Bi/multilingual – someone that uses more than one language

Call (in birds) – short, usually one-note vocalization, used in many different contexts. Some calls are referential, such as alarm call, food call

Categorical perception – the ability to discriminate between sound categories, for example between [p] and [b]

Chemical communication – communication that relies of olfaction (smell)

Chorus – vocalizations performed by many individuals simultaneously

Conspecific – of the same species

Cooperative breeding – when several individuals help with rearing the offspring

Critical period – limited time window for learning because of a high sensitivity. Also known as sensitive period or phase

Crystallized song – a stage in song development, comes after subsong and plastic song

Deictic – (From Greek *deixis* demonstration or reference). Phenomena relating to circumstances in space and time. For example finger pointing, or using words like here and there

Development of communication – how communicative behaviour develops in the young individual

Domestication – process by which animals or plants are changed at the genetic level, for human benefit. Many domesticated animals have been bred to produce artificial varieties which are not found in nature

Duet song – singing by a pair of individuals

Eavesdropping – hearing information that is not intended for the hearer

Emblem – One of the categories for the analysis of nonverbal behaviour in the model suggested by Ekman & Friesen (1969). An emblem can be translated by a word (e.g. OK-sign)

Extra-linguistic – vocalizations without language, e.g. laughter

Fast mapping – a process by which a new word can be learned with only few exposures

Focal animal sampling – an observation method by which the observer focuses on only one individual

Fission-fusion – a social group where the individuals split and merge during the day

Frequency code – a code according to which low frequency sounds are associated with large size, and high frequency sounds with small size. Suggested by Morton (1977)

Gesture – communicative movements with hands and arms. Some researchers talk about facial gestures as well

Glands – scent glands produce secretions that contain pheromones and other chemicals. They are often used for marking of territory or as sexual signals

Grammatical markers – markers of grammatical functions. For example copula verb (is), auxiliary (will, can), verb inflections (third person –s)

Homologous comparisons – comparisons between related species, for example wolves and dogs. Opposite to analoguous

Illocutionary force – the intention of the speaker. For example, if the speaker says "It is hot in here" the intent may be to ask somebody to open the window

Illustrator – one of the categories for the analysis of nonverbal behaviour in the model suggested by Ekman & Friesen (1969). An illustrator illustrates some point in the verbal utterance

Imprinting – a kind of irreversible learning early in life

Infrasound – very low frequency sound, outside the range of hearing for humans, and between 0.1 Hz and 20 Hz

Innate – from birth

Inter-specific – between species

Intra-specific – within species

Language functions – communicative functions, for example referential, social and affective. Suggested by Jakobson (1960)

Language impairment – deficits in language skills, for example grammar or pronunciation

Larynx – the vocal organ of many animals, sometimes called the "voice box". It is part of the respiratory system and contains the vocal cords, which are used in vocalizations

Mean length of Utterance (MLU) – a measure of early grammatical development, the mean value is based on how many words (or morphemes) each child utterance contains

Mirror neurons – neurons in the brain that fire both when someone observes an action and when the action is performed. This has been observed in primates and birds

Monolingual – someone that only speaks one language

Morpheme – in human language; the smallest meaningful element (e.g. dog – dogs)

Morphology
– in biology; the study of structure and form of an organism
– in linguistics; the study of morphemes

MRI – magnetic resonance imaging. A magnet field changes the molecules so that it is possible to have a three-dimensional image

Nature versus nurture – the discussion on the relative importance of innate capacities (nature) versus environmental influence (nurture)

Natural selection – Darwin suggested that animals and plants change in a non-random way

Naturalist – a scientist studying nature or natural history in general, not so specialized as a zoologist and botanist

Neonatal – newborn

New world primates – primates found in Central and South America, e.g. marmosets, tamarins, capuchins

Nonverbal – communication without words

Observer's paradox – the phenomenon where the observation of a situation is influenced by the observer. Suggested by Labov (1972)

Old world primates – primates from Africa and Asia, e.g. humans, gorillas, chimpanzees, gibbons, baboons

Olfactory communication – communication that relies on the sense of smell. So-called pheromones are not consciously observed. (See also chemical communication)

Ontogeny – development of an individual within its own lifetime as distinct from phylogeny

Oscines – "true" songbirds, a suborder of the order Passeriformes

Overextensions – when children extend the meanings of a word further than in the adult language, e.g. calling all animals 'dogs'

Parent-offspring communication – early communication between parents and infants

Paralinguistic – vocalizations together with language use, e.g. tone of voice

Passerines – a bird order that includes the suborders oscines and sub-oscines

Phoneme – speech sound, the smallest discriminating unit in a particular language (e.g. cat-rat)

Phylogeny – evolutionary history of a species. Evidence showing how different species are related to each other through evolution

Playback method – to record a behaviour (e.g. vocalization) and play it back to the group

Plastic song – an intermediate stage in bird song development, where unstable imitation and species specific song patterns first occur

Precocial – species where the young are relatively mature and mobile from birth, e.g. domestic chicken

Progressive form – special form of the verb to express on-going activity, e.g. writing, running

Recursion – the potential to create new words and sentences by combinations, e.g. embedded clauses: He said that she said that they wondered

Referential function – when an expression refers to some external phenomenon

Regulator – one of the categories for the analysis of nonverbal behaviour in the model suggested by Ekman & Friesen (1969). A regulator regulates the verbal interaction e.g. by gaze alternation

Ritualization – a behaviour that is performed in a stereotypical way, e.g. courting behaviour

Scent marking – spraying of urine or gland secretion by animals to mark their territory

Second language acquisition – the acquisition of another language after the first

Semantic feature – a meaning feature of a word. For example, the word 'woman' can be described with the features [+ adult], [+ female], and the word 'girl' with the features [– adult], [+ female]

Semantic relation – the relation between two words, e.g. in children's two-word utterances. In the example "Mummy read" there is a semantic relation between an agent (Mummy) and an action (read)

Sensitive period – see critical period

Sign language – a language using the manual signing modality

Specific Language Impairment (SLI) – a disorder in the development of language without any parallel disorder in cognitive development

Speech sounds – sounds that are used to distinguish meanings in a particular language

Spectrograph – an instrument that gives a visual representation of the acoustics of speech sounds

Syrinx – vocal organ of birds, situated further down in the trachea than the larynx. The syrinx has two different sources (at least in some birds)

Syntax – word combinations and sentence structure

Tactile signal – communicating by touch

Theory of mind – understanding the minds of others, and that others may not possess the same knowledge as you do

Triadic gesture – involving two participants and an object. For example when human children point at an object to show it to someone

Universal Grammar – a set of formal properties, assumed to be present in all human languages

Verbal – communication with words

Vocalisation – any sound produced by the vocal organs

References

Adret, P. (1993). Operant conditioning, song learning and imprinting to taped song in the zebra finch. *Animal Behaviour*, 46, 149–159.
Ahrens, R. (1954). Beitrag zur Entwicklung des Physiognomie und Mimikerkennens. *Zeitung f Experim. Angew. Psychol.* 2, 412–454, 599–633.
Albone, E. S. (1984). *Mammalian Semiochemistry. The Investigation of Chemical Signals between mammals*. New York: John Wiley & Sons.
Altmann, J. (1974). Observational study of behaviour: sampling methods. *Behaviour*, 49, 227–266.
Altmann, S. A. (Ed.). (1967). *Social Communication among Primates*. Chicago: University of Chicago Press.
Andersson, M. (1982). Female choice selects for extreme tail length in a widowbird. *Nature*, 299, 818–820.
Andersson, S. (1992). Female preference for long tails in lekking Jackson's widowbirds: experimental evidence. *Animal Behaviour*, 43, 379–388.
Arbib, M. A. (2002). Grounding the mirror system hypothesis for the evolution of the language-ready brain. In A. Cangelosi & D. Parisi (Eds.), *Simulating the evolution of language*. (pp. 229–254). London: Springer Verlag.
Arcadi, A. (1996). Phrase Structure of Wild Chimpanzee Pant Hoots: Patterns of Production and Interpopulation Variability. *American Journal of Primatology*, 39, 159–178.
Arcadi, A. (2000). Vocal responsiveness in male wild chimpanzees: implications for the evolution of language. *Journal of Human Evolution*, 39, 205–223.
Argyle, M. & Cook, M. (1976). *Gaze and mutual gaze*. Cambridge: Cambridge University Press.
Arnold, K. & Whiten A. (2003). Grooming interactions among the chimpanzees of the Budongo forest, Uganda: tests of five explanatory models. *Behaviour*, 140, 510–552.
Arnold, K. & Zuberbühler, K. (2006). Semantic combination in primate calls. *Nature*, 441, 303.
Au, W. W. L. (1993). *The Sonar of Dolphins*. New York: Springer.
Aubin, T., Jouventin, P., & Hildebrand, C. (2000). Penguins use the two-voice system to recognize each other. *Proc. Roy. Soc Lond* (Biol), 267, 1081–1087.
Austin, J. L. (1962). *How to do things with words*. London: Oxford University Press.
Baker, A., Van den Bogaerde, B. & Woll, B. (2008). Methods and procedures in sign language acquisition studies. In A. Baker, & B. Woll (Eds.), *Sign Language Acquisition*. (pp. 1–49). Amsterdam: John Benjamins.
Baker, M. C. & Cunningham, M. A. (1985). The biology of bird-song dialects. *Behavioural and Brain Sciences*, 8, 85–133.
Bangerter, A. & Oppenheimer, D. M. (2006). Accuracy in detecting referents of pointing gestures unaccompanied by language. *Gesture*, 6, 85–102.

Bard, K. A. (1998). Social-experimental contributions to imitation and emotion in chimpanzees. In S. Bråten (Ed.), *Intersubjective communication and emotion in early ontogeny: A source book.* (pp. 208–277). Cambridge, UK: Cambridge University Press.

Bard, K. A., Myowa-Yamakoshi, M., Tomonaga, M., Takaka, M., Costall, A., & Matsuzawa, T. (2005). Group differences in the mutual gaze of chimpanzees (Pan troglodytes). *Developmental Psychology,* 41, 616–624.

Barja, I., Javier de Miguel, F., & Bárcena, F. (2004). The importance of crossroads in faecal marking behaviour of the wolves (*Canis lupus*). *Naturwissenschaften,* 91, 489–492.

Barnard, P. (1990). Male tail length, sexual display intensity and female sexual response in a parasitis African finch. *Animal Behaviour,* 29, 652–656.

Bayly, K. L., Evans, C. S., & Taylor, A. (2006). Measuring social structure: A comparison of eight dominance indices. *Behavioural Processes,* 73, 1–12.

Beattie, G., & Shovelton, H. (1999a). Do iconic hand gestures really contribute anything to the semantic information conveyed by speech? An experimental investigation. *Semiotica,* 123, 1–30.

Beattie, G., & Shovelton, H. (1999b). Mapping the range of information in the iconic hand gestures that accompany spontaneous speech. *Journal of Language and Social Psychology,* 18, 438–462.

Beattie, G., & Shovelton, H. (2002). What properties of talk are associated with the generation of spontaneous iconic gestures? *British Journal of Social Psychology,* 41, 403–417.

Beecher, M. D. (1996). Birdsong learning in the Laboratory and Field. In D. E. Kroodsma, & E. H. Miller (Eds.), *Ecology and evolution of acoustic communication in birds.* (pp. 61–78). Ithaca, N.Y.: Cornell University Press.

Beecher, M. D., & Brenowitz, E. A. (2005). Functional aspects of song learning in songbirds. *Trends in Ecology and Variation,* 20, 143–149.

Beecher, M. D., Stoddard, P. K., & Loesche, P. (1985). Recognition of Parents' Voices by Young Cliff Swallows. *The Auk,* 102, 600–605.

Bekoff, M. (2001). Observations of scent-marking and discriminating self from others by a domestic dog (*Canis familiaris*): tales of displaced yellow snow. *Behavioural Processes,* 55, 75–79.

Beneventi, H., Barndon, R., Ersland, L., & Hugdahl, K. (2007). An fMRI study of working memory for schematic facial expressions. *Scandinavian Journal of Psychology,* 48, 81–86.

Berko, J. (1958). The child's learning of English morphology. *Word,* 14, 150–177.

Berko Gleason, J. (1977). Talking to children: some notes on feedback. In Snow, C. & Ferguson, C. (Eds.), *Talking to children. Language Input & Acquisition.* (pp. 199–205). Cambridge: Cambridge University Press.

Best, C. T., McRoberts, G. W., LaFluer, R., & Silver-Isenstadt, J. (1995). Divergent developmental patterns in infants' perception of two non-native consonant contrasts. *Infant Behavior & Development,* 19, 339–350.

Bickerton, D. (1995). *Language and Human Behaviour.* Seattle: University of Washington Press.

Bingham, H. C. (1927). Parental play of chimpanzee. *Journal of Mammalogy,* 8, 77–89.

Bishop, D. V. M. (2006). What Causes Specific Language impairment in Children? *Current Directions in Psychological Science,* 15, 217–221.

Bishop, D. V. M., North, T., & Donlan C. (1995). Genetic basis of specific Language Impairment: evidence from a twin study. *Developmental Medicine and Child Neurology,* 37, 56–71.

Bley-Vroman, R. (1989). What is the logical problem of foreign language learning? In S. M. Gass, & J. Schachter (Eds.), *Linguistic Perspectives on Second Language Acquisition.* (pp. 41–68). Cambridge: Cambridge University Press.

Blomqvist, C. (2004). Directional aggressive pulse sounds in the Bottlenose Dolphins (*Tursiops truncates*): technical aspects and social implications. *Linköping Studies in Science and Technology*. Dissertation no 905. University of Linköping, Sweden.

Blomqvist, C., & Amundin, M. (2004). High Frequency Burst-Pulse Sounds in Agonistic/Aggressive Interactions in Bottlenose Dolphins, *Tursiops truncates*. In J. A. Tomas, C. F. Moss, & M. Vaters (Eds.), *Echolocation in Bats and Dolphins*. (pp. 425–431). Chicago: University of Chicago Press.

Blumstein, D. T., & Armitage, K. B. (1997). Alarm calling in yellow-bellied marmots: I The meaning of situationally variable alarm calls. *Animal Behaviour*, 53, 143–171.

Boesch, C., & Boesch-Achermann, H. (2000). *The Chimpanzees of the Taï Forest. Behavioural Ecology and Evolution*. Oxford: Oxford University Press.

Boitani, L., Ciucci, P., & Ortolani, A. (2007). Behaviour and Social Ecology of Free-ranging Dogs. In P. Jensen (Ed.), *The behavioural biology of dogs*. (pp. 147–165). Wallingford, Oxfordshire: CABI International Press.

Bostwick, K. S., & Prum, R. O. (2005). Courting bird sings with stridulating wing feathers. *Science*, 309, 736.

Boughman, J. W., & Moss, C. F. (2003). Social Sounds: Vocal Learning and Development of Mammal and Bird Calls. In A. M. Simmons, A. N. Popper, & F. F. Rickard (Eds.), *Acoustic Communication*. (pp. 138–224). New York: Springer

Bowerman, M. (1978). Systematizing Semantic Knowledge: Changes over Time in the Child's Organization of Word meaning. *Child Development*, 49, 977–987.

Bowerman, M. (1994). From Universal to Language-Specific in Early Grammatical Development. *Philosophical Transactions: Biological Sciences*, 346(1315), 37–45.

Bowerman, M., & Choi, S. (2003). Space under construction: Language-specific spatial categorization in first language. In D. Gentner, & S. Goldin-Meadow (Eds.), *Language in mind: Advances in the study of language and thought*. (pp. 388–427). Cambridge, Mass.: MIT Press.

Bowlby, J. (1973). *Attachment and Loss: Separation, Anxiety, and Anger*. New York: Basic Books.

Boysson-Bardies, B. de, Hallé, P., Sagart, L., & Durand, C. (1989). A crosslinguistic investigation of vowel formants in babbling. *Journal of Child Language*, 16, 1–17.

Bradbury, J. W., & Vehrencamp, S. L. (1998). *Principles of Animal Communication*. Sunderland MA: Sinauer.

Bråten, S. (2002). Altercentric perception by infants and adults in dialogue. Ego's virtual participation in Alter's complementary act. In M. I. Stamenov, & V. Gallese (Eds.), *Mirror Neurons and the Evolution of Brain and Language*. (pp. 273–294). Amsterdam: John Benjamins.

Brennan, P. A., & Keverne, E. B. (2004). Something in the air? New insights into mammalian pheromones. *Current Biology*, 14, 81–89.

Bretagnolle, V. (1996). Acoustic Communication in a Group of Nonpasserine Birds, the Petrels. In D. E. Kroodsma, & E. H. Miller (Eds.), *Ecology and Evolution of Acoustic Communication in Birds*. (pp. 160–177). Ithaca, N.Y.: Cornell University Press.

Brinck, I., & Gärdenfors, P. (2003). Co-operation and Communication in Apes and Humans. *Mind & Language*, 18, 484–501.

Brosnahan, I. T. (1979). The haptics of the English handshake. *IRAL XVII*, 77–82.

Brown, R. (1973). *A first language. The early stages*. Cambridge, Mass. Harvard University Press.

Brown, C., Bomberger Brown, M., & Shaffer, M. L. (1991). Food-sharing signals among socially foraging cliff swallows. *Animal Behaviour*, 42, 551–564.

Brown, P. (2001). Learning to talk about motion UP and DOWN in Tzeltal: is there a language-specific bias for verb learning? In M. Bowerman, & S. C. Levinson (Eds.), *Language acquisition and conceptual development*. (pp. 512–543). Cambridge: Cambridge University Press.

Buck, L. B. (2000). The molecular architecture of odor and pheromone sensing in mammals. *Cell*, 100, 611–618.

Buck, L. B., & Axel, R. (1991). A novel multigene family may encode odorant receptors – a molecular-basis for odor recognition. *Cell*, 65, 175–187.

Buchanan, K. L., Catchpole, C. K., Lewis, J. W., & Lodge, A. (1999). Song as an indicator of parasitism in the sedge warbler. *Animal Behaviour*, 57, 307–314.

Burling, R. (2005). *The Talking Ape*. Oxford: Oxford University Press.

Burnham, D., Kitamura, C., & Vollmer-Conna, U. (2002). What's New Pussycat? On Talking to Babies and Animals. *Science*, 296, 1435.

Busch, R. H. (1995). *The Wolf Almanac*. New York: Lyons & Burford.

Burt, J. M., & Vehrencamp, S. (2005). Dawn chorus as an interactive communication network. In P. McGregor (Ed.), *Animal Communication Networks*. (pp. 320–343). Cambridge: Cambridge University Press.

Butterworth, G. E. (1995). Factors in Visual Attention Eliciting Manual Pointing in Human Infancy. In H. L. Roitblat, & J-A. Meyer (Eds.), *Comparative Approaches to Cognitive Science*. (pp. 329–338). Cambridge, Mass.: MIT Press.

Byers, B. E., & Kroodsma, D. E. (1992). Development of two song categories by chestnut-sided warblers. *Animal Behaviour*, 44, 799–810.

Calbris, G. (1990). *The Semiotics of French Gestures*. Bloomington: University of Indiana Press.

Calbris, G. (2011). *Elements of Meaning in Gesture*. Amsterdam: John Benjamins.

Caldwell, M. C., & Caldwell, D. K. (1990). Review of the signature whistle hypothesis for the Atlantic bottlenose dolphin. In S. Leatherwood, & R. R. Reeves (Eds.), *The Bottlenose dolphins (Tursiops truncates)*. (pp. 199–234). San Diego: Academic Press.

Call, J., Hare, B. A., & Tomasello, M. (1998). Chimpanzee gaze following in an object-choice task. *Animal Cognition*, 1, 89–99.

Call, J., & Tomasello, M. (2008). Does the chimpanzee have a theory of mind? 30 years later. *Trends in Cognitive Science*, 12, 187–192.

Candland, D. K., & Bush, S. L. (1995). Primates and Behavior. In E. Gibbons, B. Durrant, & J. Demarest (Eds.), *Conservation of Endangered Species in Captivity*. (pp. 521–551). New York: State University of New York Press.

Cangelosi, A., & Parisi, D. (Eds.). (2002). *Simulating the evolution of language* London: Springer Verlag.

Carpenter, M., Nagell, K., & Tomasello, M. (1998). Social cognition, joint attention, and communicative competence from 9 to 15 months of age. *Monographs of the Society of Research in Child Development*, 63(4).

Catchpole, C. K. (1976). Temporal and sequential organisation of song in the sedge warbler (Acrocephalus schoenobaenus). *Behaviour*, 59, 226–246.

Catchpole, C. K., & Slater, P. J. B. (1995). *Bird song. Biological themes and variations*. Cambridge: Cambridge University Press.

Cheney, D. L., & Seyfarth, R. M. (1980). Vocal recognition in free-ranging vervet monkeys. *Animal Behaviour*, 28, 362–367.

Cheney, D. L., & Seyfarth, R. M. (1982). How vervet monkeys perceive their grunts: field playback experiments. *Animal Behaviour*, 30, 739–751.

Cheney, D. L., & Seyfarth, R. M. (1990). *How monkeys see the world*. Chicago: University of Chicago Press.
Cheyne, S. M., Chivers, D. J., & Sugardjito, J. (2007). Covariation in the great calls of rehabilitant and wild gibbons (*Hylobates albibarbis*). *The Raffles Bulltin of Zoology*, 55, 201-207.
Chomsky, N. (1957). *Syntactic Structures*. The Hague: Mouton.
Chomsky, N. (1965). *Aspects of the Theory of Syntax*. Cambridge. Mass.: MIT Press.
Chomsky, N. (1980). *Rules and Representations*. New York: Columbia Press.
Chomsky, N. (2000a). *The Architecture of Language*. New Delhi: Oxford University Press.
Chomsky, N. (2000b). *New Horizons in the Study of language and Mind*. Cambridge: Cambridge University Press.
Chomsky, N. (2004). Language and Mind: Current thoughts on Ancient Problems. In L. Jenkins (Ed.), *Variation and Universals in Biolinguistics*. (pp. 379-405). Amsterdam: Elsevier.
Christiansen, M. H., & Kirby, S. (Eds.). (2003). *Language Evolution*. Oxford: Oxford University Press.
Christoffersson, S., & Bärg, U. (1990). Fågelhundar: dressyr och jakt. [The setting dog, education and hunting]. Västerås: ICA-förlaget.
Clark, A. P., & Wrangham, R. W. (1994). Chimpanzee arrival pant hoots: Do they signify food or status? *International Journal of Primatology*, 15, 185-205.
Clark, E. V. (1978). Strategies for Communicating. *Child Development*, 49, 953-959.
Clark, E. V. (1993). *The lexicon in acquisition*. Cambridge: Cambridge University Press.
Cockburn, A. (2006). Prevalence of different modes of parental care in birds. *Proc R. Soc. Lond. B.*, 273, 1375-1383.
Collias, N. E. (1960). An Ecological and Functional classification of Animal Sounds. In W. E. Lanyon & W. N. Tavolga (Eds.), *Animal sounds and communication*. (pp. 368-391). Washington D.C.: American Institute of Biological Sciences.
Collias, N. E. (1987). The vocal repertoire of the red junglefowl: A spectrographic classification and the code of communication. *Condor*, 89, 510-524.
Collins, S. (2004). Vocal fighting and flirting: the functions of birdsong. In P. Marler & H. Slabbekoorn (Eds.), *Nature's Music* (pp. 39-79). London: Elsevier.
Condon, W. S., & Sander, L. W. (1974). Neonate Movement Is Synchronized with adult Speech: Interactional Participation and Language Acquisition. *Science*, 183, 99-101.
Cools, A. K. A., Van Hout, A. J.-M., & Nelissen, M. H. J. (2008). Canine Reconciliation and Third-Party-Initiated Postconflict Affiliation: Do Peacemaking Social Mechanisms in Dogs Rival Those of Higher Primates? *Ethology*, 114, 53-63.
Coppinger, R., & Coppinger, L. (2001). *Dogs. A New Understanding of Canine Origin, Behavior, and Evolution*. Chicago: University of Chicago press.
Coppinger, R., & Schneider, R. (1995). Evolution of working dogs. In J. Serpell (Ed.), *The Domestic Dog: Its Evolution, Behaviour and Interactions with People*. (pp. 21-47). Cambridge: Cambridge University Press.
Cordoni, G., & Palagi, E. (2008). Reconciliation in Wolves (*Canis lupus*): New Evidence for a Comparative Perspective. *Ethology*, 114, 298-308.
Coropassi, K. A., & Bradbury, J. W. (2006). Contact call diversity in wild orange-fronted parakeet pairs. *Aratinga canicularis. Animal Behaviour*, 71, 1141-1154.
Costa, A., Pickering, M. J., & Sorace, A. (2008). Alignment in second language dialogues. *Language and Cognitive Processes*, 23, 528-556.
Coulson, M. (2004). Attributing emotion to static body postures: recognition and accuracy, confusions, and viewpoint dependence. *Journal of Nonverbal behavior*, 28, 117-139.

Coward, S. W., & Stevens, C. J. (2004). Extracting meaning from sound: nomic mappings, everyday listening, and perceiving object size from frequency. *The Psychological Record,* 54, 349–364.

Crago, M. (1992). Ethnography and language socialization: A cross-cultural perspective. *Topics in Language Disorders,* 12, 28–39.

Crisler, L. (1959). *Arctic wild.* New York: Harper and Row.

Crystal, D. (2006). *How language works: how babies babble, words change their meaning and languages live or die.* London: Penguin

Crockford, C., & Boesch, C. (2003). Context-specific calls in wild chimpanzees, Pan troglodytes verus: analysis of barks. *Animal Behaviour,* 66, 115–125.

Crockford, C., & Boesch, C. (2005). Call combinations in wild chimpanzees. *Behaviour,* 142, 397–421.

Crockford, C., Herbinger, I., Vigilant, L., & Boesch, C. (2004). Wild Chimpanzee Produce Group-Specific Calls: a Case for Vocal Learning? *Ethology,* 110, 221–243.

Curtis, C. C., & Stoddard, P. (2003). Mate preference in female electric fish, Brachyhypopomus pinnicaudatus. *Animal Behaviour,* 66, 329–336.

Curtiss, S. (1977). *Genie: A linguistic study of a modern-day "wild child".* New York: Academic Press.

Cuthill, I. C., & MacDonald, W. A. (1990). Experimental manipulation of the dawn and dusk chorus in the blackbird (Turdus merula). *Behav. Ecol. Sociobiol.* 26, 209–216.

Cuthill, I. C., Partridge, J. C., & Bennett, A. T. D. (2000). Avian UV Vision and Sexual Selection. In Y. Espmark, T. Amundsen, & G. Rosenqvist (Eds.), *Animal Signals: Signalling and Signal Design in Animal Communication.* (pp. 61–82). Trondheim: Tapir Academic Press.

Cynx, J., & Clark, S. (1998). The laboratory use of conditional and natural responses in the study of avian auditory perception. In S. L. Hopp, M. J. Owren, & C. S. Evans (Eds.), *Animal Acoustic Communication.* (pp. 353–377). Berlin: Springer Verlag.

Cynx, J., Leweis, R., Tavel, B., & Hanson, T. (1998). Amplitude regulation of vocalizations in noise by a songbird, Taeniopygia guttata. *Animal Behaviour,* 56, 107–113.

Dabelsteen, T., & McGregor, P. K. (1996). Dynamic acoustic communication and interactive playback. In Kroodsma, D. E. & Miller, E. H. (Eds.), *Ecology and Evolution of Acoustic Communication in Birds.* (pp. 398–408). Ithaca: Comstock.

Darwin, C. (1872/1965). *The expression of the emotions in man and animals.* Chicago: University of Chicago Press.

Darwin, C. (1877). A biographical sketch of an infant. *Mind. A Quarterly Review of Psychology and Philosophy,* 7, 285–294.

Davidson, I. (1999). The game of the name: Continuity and discontinuity in language origins. In B. J. King (Ed.), *The origins of language: What nonhuman primates can tell us.* (pp. 229–268). Santa Fe: N.M. School of American Research Press.

Davies, N. B., & Lundberg, A. (1984). Food distribution and a variable mating system in the dunnock (Prunella modularis). *Ibis,* 127, 100–110.

Davis, B. L. (2011). Illuminating language origins from the perspective of contemporary ontogeny in human infants. In A. Valain, J.-L. Schwartz, C. Abry, & J. Vauclair (Eds.), *Primate Communication and Human Language.* (pp. 173–192). Amsterdam: John Benjamins.

de Boer, B. (2005). Infant-directed speech and evolution of language. In M. Tallerman (Ed.), *Language Origins: Perspectives on Evolution.* (pp. 100–121). Oxford: Oxford University Press.

DeCasper, A. J., Leacanuet, J-P., Busnel, M-C., Granier-Deferre, C., & Maugeais, R. (1994). Fetal Reactions to Recurrent Maternal Speech. *Infant Behavior and Development,* 17, 159–164.

de Waal, F. B. M. (1994). Chimpanzee's Adaptive Potential. A Comparison of Social Life under Captive and Wild Conditions. In R. W. Wrangham, W. C McGrew, F. B. M. de Waal, & P. G. Heltne (Eds.), *Chimpanzee Cultures*. (pp. 243–260). Harvard University Press.

de Waal, F. B. M. (2001). *The Ape and the Sushi Master: Cultural Reflections by a Primatologist*. New York: Basic Books.

de Waal, F. B. M., & Lanting, F. (1997). *Bonobo. The forgotten ape*. Berkeley: University of California Press.

DeVoogd, T. J., & Székely, T. (1998). Causes of avian song: using neurobiology to integrate proximate and ultimate levels of analysis. In R. P. Balda, & I. M. Pepperberg (Eds.), *Animal Cognition in Nature*. (pp. 337–380). San Diego: Academic Press.

DeVore, I. (Ed.). (1968). *Primate Behavior: Field Studies of Monkeys and Apes*. New York: Holt, Rinehart & Winston.

DeVore, I. (1973). Primate Behavior. In M. Argyle (Ed.), *Social Encounters. Readings in Social Interaction*. (pp. 17–33). Chicago: Aldine Pub. Co.

Deacon, T. W. (1997). *The symbolic species. The Co-evolution of language and the brain*. London: W. W. Norton & Company.

Di Bitetti, M. S. (2005). Food-associated calls and audience effects in tufted capuchin monkeys *Cebus apella nigritus*. *Animal Behaviour*, 69, 911–919.

di Pellegrino, G., Fadiga, L., Fogassi, L., Gallese, V., & Rizzolatti, G. (1992). Understanding motor events: a neurophysiological study. *Experimental Brain Research*, 91, 176–180.

Digweed, S. M., Fedigan, L. M., & Rendall, D. (2005). Variable specificity in the anti-predator vocalizations and behaviour of the white-faced capuchin, Cebus capicinus. *Behaviour*, 142, 1003–1027.

Dimberg, U., Thunberg, M., & Elmehed, K. (2000). Unconscious facial reactions to emotional facial expressions. *Psychological Science*, 11, 86–89.

Van Dongen, W. F. D. (2006). Variation in singing behaviour reveals possible function of song in male golden whistlers. *Behaviour*, 143, 57–82.

Dooling, R. J., Best, C. T., & Brown, S. D. (1995). Discrimination of synthetic full-formant and sinewave /ra-la/ continua by budgerigars (*Melopsittacus undulates*) and zebra finches. *Journal of the Acoustic Society of America*, 97, 1839–1846.

Dorries, K. M., Adkins-Regan, E., & Halpern, B. P. (1997). Sensitivity and behavioural responses to the pheromone androstenone are not mediated by the vomeronasal organ in domestic pigs. *Brain Behav. Evol.* 49, 53–62.

Duchenne de Boulogne, G-B. (1862/1990). *The Mechanism of Human Facial Expression*. Cambridge: Cambridge University Press.

Dunbar, R. I. M. (1991). Functional significance of social grooming in primates. *Folia Primatol.* 57, 121–131.

Dunbar, R. I. M. (1996). *Grooming, Gossip and the Evolution of Language*. London: Faber & Faber.

Eens, M., Pinxten, R., & Verheyen, R. F. (1992). Song learning in captive European starlings, *Sturnus vulgaris*. *Animal Behaviour*, 44, 1131–1143.

Efron, D. (1941). *Gesture and Environment*. New York: King's Crown.

Eibl-Eibesfeldt, I. (1970). *Ethology. The biology of behavior*. New York: Holt, Rinehart & Winston.

Eibl-Eibesfeldt, I. (1972). Similarities and differences between cultures in expressive movements. In R. A. Hinde (Ed.), *Non-verbal Communication*. (pp. 297–312). Cambridge: Cambridge University Press.

Eibl-Eibesfeldt, I. (1986). *Human Ethology*. New York: de Gruyter.

Eimas, P. D., Siqueland, E. R., Juscyk, P. W., & Vigorito, J. (1971). Speech perception in infants. *Science*, 171, 303–306.

Ekman, J., & Ericson, P. G. P. (2006). Out of Gondwanaland; the evolutionary history of cooperative breeding and social behaviour among crows, magpies, jays and allies. *Proceedings of the Royal Society B*, 273, 1117–1125.

Ekman P., & Friesen, W. V. (1969). The Repertoire of Nonverbal Behavior: Categories, Origins, Usage, and Coding. *Semiotica*, 1, 49–98.

Ekman, P., & Friesen, W. V. (1978). *Facial action coding system*. Palo Alto, CA: Consulting Psychologists Press.

Elgar, M. A. (1986). House sparrows establish foraging flocks by giving chirrup calls if the resources are divisible. *Animal Behaviour*, 34, 169–174.

Elias, D. O., Hebets, E. A., Hoy, R. R., & Mason, A. C. (2005). Seismic signals are crucial for male mating success in a visual specialist jumping spider (*Araneae: Salticidae*). *Animal Behaviour*, 69, 931–938.

Ellis, R., & Larsen-Freeman, D. (2006). Language emergence: Implications for applied linguistics – Introduction to the special issue. *Applied Linguistics*, 27, 558–589.

Elowson, M. A. & Snowdon, C. T. (1994). Pygmy marmosets, *Cebuella pygmaea*, modify vocal structure in response to changed social environment. *Animal Behaviour*, 47, 1267–1277.

Elowson, M. A., Snowdon, C. T., & Lazaro-Perea, C. (1998). "Babbling" and social context in infant monkeys: parallels to human infants. *Trends in cognitive Sciences*, 2, 31–37.

Emery, N. J. (2000). The eyes have it: The neuroethology, function and evolution of social gaze. *Neuroscience and Biobehavioural Reviews*, 24, 581–604.

Enfield, N. J., & Levinson, S. C. (Eds.). (2006). *Roots of Human Sociality. Culture, Cognition and Interaction*. Oxford: Berg.

Eriksen, N., Miller, L. A., Tougaard, J., & Helweg, D. A. (2005). Cultural change in the songs of humpback whales (Megaptera novaeangliae) from Tonga. *Behaviour*, 142, 305–328.

Evans, C. S. (1991). Of ducklings and Turing machines: Interactive playbacks enhance subsequent responsiveness to conspecific calls. *Ethology*, 89, 125–134.

Evans, C. S., & Evans, L. (1999). Chicken food calls are functionally referential. *Animal Behaviour*, 58, 307–319.

Evans, C. S., & Evans, L. (2007). Representational signalling in birds. *Biology Letters*, 3, 8–11.

Evans, C. S., Evans, L., & Marler, P. (1993). On the meaning of alarm calls: Functional reference in an avian vocal system. *Animal Behaviour*, 46, 23–38.

Evans, C. S., Macedonia, J. M., & Marler, P. (1993). Effects of apparent size and speed on the response of chickens, *Gallus gallus*, to compter-generated simulations of aerial predators. *Animal Behaviour*, 46, 1–11.

Evans, C. S., & Marler, P. (1991). On the use of video images as social stimuli in birds: Audience effects on alarm calling. *Animal Behaviour*, 41, 17–26.

Evans, C. S., & Marler, P. (1994). Food calling and audience effects in male chickens, Gallus gallus: their relationships to food availability, courtship and social facilitation. *Animal Behaviour*, 47, 1159–1170.

Evans, C. S., & Marler, P. (1995). Language and Animal Communication: Parallels and Contrasts. In H. L. Roitblat, & J.-A. Meyer (Eds.), *Comparative Approaches to Cognitive Science*. (pp. 341–382). Cambridge, Mass.: MIT Press.

Evans, N. (2003). Context, culture and structuration in the languages of Australia. *Annual Review of Anthropology*, 32, 13–40.

Evans, N. (2010). *Dying words. Endangered languages and what they have to tell us.* Malden, MA.: Wiley-Blackwell.

Everett, D. (2005). Cultural Constraints on Grammar and Cognition in Pirahã: Another Look at the Design Features of Human Language. *Current Anthropology*, 76/4, 621–646.

Fadiga, L., Craighero, L., Buccino, G., & Rizzolatti, G. (2002). Speech listening specifically modulates the excitability of tongue muscles: a TMS study. *European Journal of Neuroscience*, 15, 399–402.

Farabaugh, S. M. (1982). Ecological and Social Significance of Duetting. In D. E. Kroodsma, & E. H. Miller (Eds.), *Acoustic Communication in Birds,* Vol 2. (pp. 85–124). New York: Academic Press.

Farabaugh. S. M., & Dooling, R. J. (1996). Acoustic communication in Parrots: Laboratory and Field Studies of Budgerigars, Melopsittacus undulates. In D. E. Kroodsma, & E. H. Miller (Eds.), *Ecology and Evolution of Acoustic Communication in Birds.* (pp. 97–117). Ithaca, N.Y.: Cornell University press.

Feddersen-Petersen, D. (1991). The ontogeny of social play and agonistic behaviour in selected canid species. *Bonner Zoologische Beiträge*, 42, 97–114.

Feddersen-Petersen, D. U. (2007). Social Behavior of Dogs and Related Canids. In P. Jensen (Ed.), *The behavioural biology of dogs.* (pp. 105–119). Wallingford, Oxfordshire: CABI International Press.

Feh, C., & Mazières, J. de. (1993). Grooming at preferred site reduces heart rate in horses. *Animal Behaviour*, 46, 1191–1194.

Feinberg, D. R., Jones, B. C., Little, A. C., Burt, D. M., & Perrett, D. I. (2005). Manipulations of fundamental and formant frequencies influence the attractiveness of human male voices. *Animal Behaviour*, 69, 561–568.

Ferguson, C. (1977). Baby talk as a simplified register. In C. E. Snow & C. A. Ferguson (Eds.), *Talking to children.* (pp. 219–235). Cambridge: Cambridge University Press.

Fichtel, C. (2004). Reciprocal recognition of sifaka (*Propithecus verreauxi verreauxi*) and the red-fronted lemur (*Eulemur fulvus rufus*) alarm calls. *Animal Cognition*, 7, 45–52.

Fink, B., Grammer, K., & Matts, P. J. (2006). Visible skin color distribution plays a role in the perception of age, attractiveness, and health in female faces. *Evolution & Human Behavior*, 27, 6, 433–442.

Fisher, S. E. (2005). Dissection of molecular mechanisms underlying speech and language disorders. *Applied Psycholinguistics*, 26, 111–128.

Fisher, J., Metz, M., Cheney, D. L., & Seyfarth, R. M. (2001). Baboon responses to graded bark variants. *Animal Behaviour*, 61, 925–931.

Fitch, W. T. (2002). Comparative Vocal Production and the Evolution of Speech: Reinterpreting the Descent of the Larynx. In A. Wray (Ed.), *The Transition to Language.* (pp. 21–45). Oxford: Oxford University Press.

Fitch, W. T. (2010). *The Evolution of Language.* Cambridge: Cambridge University Press.

Ford, J. K. B., & Fisher, H. D. (1983). Group-specific dialects of Killer whales (*Orcinus orca*) in British Columbia. In R. Payne (Ed.), *Communication & Behaviour of Whales.* (pp. 129–161). Boulder. Colorado: Westview Press.

Forkman, B. (2000). Domestic hens have declarative representations. *Animal Cognition*, 3, 135–137.

Fossey, D. (1983). *Gorillas in the Mist.* Boston: Houghton Mifflin.

Fox, N., & Davidson, R. (1988). Patterns of brain electrical activity during facial signs of emotion in 10 month old infants. *Developmental Psychology*, 24, 230–236.

Franz, C. (1999). Allogrooming Behavior and Grooming Site Preferences in Captive Bonobos (Pan Paniscus): Association with female dominance. *International Journal of Primatology*, 20, 525–546.

Freeberg, T. M., Dunbar, R. I., & Ord, T. J. (2012). Social complexity as a proximate and unltimate factor in communicative complexity. *Philosophical transactions of the Royal Society B*, 367, 1785-1801.

Frisch, K. von. (1954). *The Dancing Bees*. London: Methuen & Co.

Frisch, K. von. (1967). *The Dance Language and Orientation of Bees*. Mass.: Harvard University Press.

Fukuzawa, M., Mills, D. S., & Cooper, J. J. (2005). The effect of human command phonetic characteristics on auditory cognition in dogs. *Journal of Comparative Psychology*, 119, 117-120.

Gácsi, M., Györi, B., Miklósi, A., Viranyi, Zs., Kubinyi, E., Topál, J., & Csányi, V. (2005). Species-specific Differences and Similarities in the Behaviour of Hand-Raised Dog and Wolf in Social Situations with Humans. *Developmental Psychobiology*, 27, 111-122.

Gallese, V., Fadiga, L., Fogassi, L., & Rizzolatti, G. (1996). Action recognition in the premotor cortex. *Brain*, 119, 593-609.

Gallistel, C. R. (1990). Representations in animal cognition: An introduction. *Cognition*, 37, 1-22.

Gammon, D. E., & Baker, M. C. (2004). Song repertoire evolution and acoustic divergence in a population of black-capped chickadees, *Poecile atricapillus*. *Animal Behaviour*, 68, 903-913.

Gärdenfors, P. (1996). Cued and detached representations in animal cognition. *Behavioural Processes*, 35, 263-273.

Gärdenfors, P. (2004). Cooperation and the Evolution of Symbolic Communication. In D. Oller & U. Griebel, U. (Eds.), *The Evolution of Communication Systems: A Comparative Approach*. (pp. 237-256). The Vienna Series in Theoretical Biology, Cambridge MA.: MIT Press.

Gardner, R. A., & Gardner, B. T. (1969). Teaching sign language to a chimpanzee. *Science*, 165, 664-672.

Gardner, R. A., & Gardner, B. T. (1984). A Vocabulary Test for Chimpanzees (*Pan troglodytes*). *Journal of Comparative Psychology*, 98, 381-404.

Garnica, O. (1977). Some prosodic and paralinguistic features of speech to young children. In C. Snow, & C. Ferguson (Eds.), *Talking to children. Language Input & Acquisition*. (pp. 63-88). Cambridge: Cambridge University Press.

Gaskins, S. (2006). Cultural Perspectives on Infant-Caregiver Interaction. In N. J. Enfield, & S. C. Levinson (Eds.), *Roots of Human Sociality. Culture, Cognition and Interaction*. (pp. 279-298). Oxford: Berg.

Gathercole, S. E. (2006). Nonword repetition and word learning: The nature of the relationship. *Applied Psycholinguistcs*, 27, 513-543.

Gaunet, F. (2008). How do guide dogs of blind owners and pet dogs of sighted owners (*Canis familiaris*) ask their owners for food? *Animal Cognition*, 11, 475-483.

Geissmann, T. (1983). Female capped gibbon (Hylobates piteatus Gray 1981) sings male song. *Journal of Human Evolution*, 12, 667-671.

Geissmann, T. (2000). Gibbon songs and human music from an evolutionary perspective. In N. L. Wallin, B. Merker & S. Brown (Eds.), *The origins of Music*. (pp. 103-124). Cambridge, Mass.: MIT Press.

Geissmann, T. (2002). Duet-splitting and the evolution of gibbon song. *Biological Review*, 77, 57-76.

Geissmann, T. (2003). Tape-recording primate vocalisations. In J. M Setchell, & D. J. Curtis (Eds.), *Field and Laboratory Methods in Primatology: A Practical Guide*. (pp. 228-238). Cambridge: Cambridge University Press.

Gentner, T. Q., Fenn, K. M., Margoliash, D., & Nusbaum, H. C. (2006). Recursive syntactic pattern learning by songbirds. *Nature*, 440, 1204-1207. (27 April 2006).

Ghiglieri, M. P. (1984). *The Chimpanzees of Kibale Forest: A Field Study of Ecology and Social Structure*. New York: Columbia University Press.

Giles, H. (1980). Accommodation theory: some new directions. In S. de Silva (Ed.), *Aspects of linguistic behaviour* (pp. 105–136). York, UK: York University Press.

Giles, H., Taylor, D. M., & Bourhis, R. (1973). Towards a theory of interpersonal accommodation through language: some Canadian data. *Language in Society*, 2, 177–192.

Goffman, E. (1959). *The presentation of self in everyday life*. New York: Doubleday Anchor.

Goffman, E. (1963). *Behaviour in public places*. New York: The Free Press.

Goffman, E. (1967). *Interaction Ritual. Essays on Face-to Face Behaviour*. New York: Anchor Books.

Goldin-Meadow, S., & Butcher, C. (2003). Pointing toward two-word speech in young children. In S. Kita (Ed.), *Pointing: Where language, culture, and cognition meet*. (pp. 85–107). Mahwah, NJ: Earlbaum Associates.

Goodall, J. (1965). Chimpanzees of the Gombe Stream Reserve. In I. DeVore (Ed.), *Primate behaviour. Field Studies of Monkeys and Apes*. (pp. 425–447). New York: Holt Rinehart & Winston.

Goodall, J. (1967). Mother-offspring relationships in free-ranging chimpanzees. In D. Morris (Ed.), *Primate ethology*. (pp. 287–346). London: Wedenfeld & Nicolson.

Goodall, J. (1968). Expressive movements and communication in free-ranging chimpanzees: a preliminary report. In P. Jay (Ed.), *Primates: Studies in Adaptation and Variability*. (pp. 313–374). New York: Holt, Rinehart & Winston.

Goodall, J. (1986). *The chimpanzees of Gombe, patterns of behaviour*. Cambridge, England: Belknap Press of Harvard University Press.

Goodall, J. (2005). Great ape biology. In J. Caldecott & I. Miles (Eds.), *World Atlas of Great Apes and their conservation*. (pp. 29–30). Berkeley: University of California Press.

Goodwin, D., Bradshaw, J. W. S., & Wickens, S. M. (1997). Paedomorphosis affects agonistic signals of domestic dogs. *Animal Behaviour*, 53, 297–304.

Gopnik, M. (1990). Featureblind grammar and dysphasia. *Nature*, 344, 715.

Gould, J. L. (1990). Honey bee cognition. *Cognition*, 37, 83–102.

Gould, J. L., & Gould Grant, C. (1994). *The animal mind*. New York: Scientific American Library.

Granström, K. (1992). *Dominans och underkastelse hos tonårspojkar. En studie av icke-verbala kommunikationsmönster*. [Dominance and subordination in adolescents. A study of non-verbal communicative patterns]. *SIC* 34. University of Linköping.

Grassman, S., Kaminski, J., & Tomasello, M. (2012). How two word-trained dogs integrate pointing and naming. *Animal Cognition*, 15, 657–665.

Greenbaum, P. E., & Rosenfeld, H. M. (1980). Varieties of touching in greetings: Sequential structure and sex-related differences. *Journal of Nonverbal Behavior*, 5, 13–25.

Greene, E., & Meagher, T. (1998). Red squirrels, *Tamiasciurus hudsonicus*, produce predator-class specific alarm calls. *Animal Behaviour*, 55, 511–518.

Griesser, M. (2009). Referential calls signal predator behavior in a group-living bird species. *Current Biology*, 18, 69–73.

Gros-Louis, J. (2004). The function of food-associated calls in white-faced capuchin monkeys, Cebus capucinus, from the perspective of the signaller. *Animal Behaviour*, 67, 431–440.

Gullberg, M. (1998). *Gesture as a communication strategy in second language discourse*. Lund University Press.

Gullberg, M. (2006). Handling Discourse: Gestures, Reference Tracking, and Communication Strategies in Early L2. *Language Learning*, 56, 155–196.

Gullberg, M. & Holmqvist, K. (2006). What speakers do and what addressees look at. *Pragmatics & Cognition*, 14, 53–82.

Györi, G. (1998). Cultural variation in the conceptualisation of emotions. In A. Athanasiadou, & E. Tabakowska (Eds.), *Speaking of Emotions. Conceptualisation and Expression.* (pp. 99–125). Berlin: Mouton de Gruyter.

Haesler, S., Wada, K., Nshdejan, A., Morrisey, E. E., Lints, T., Jarvis, E. J., & Scharff, C. (2004). FoxP2 expression in avian vocal learners and non-learners. *Journal of Neuroscience*, 24, 3164–3175.

Håkansson, G. (1987). *Teacher Talk. How teachers modify their language when addressing learners of Swedish as a second language.* Lund: Lund University Press.

Håkansson, G. (1989). Bodily Behaviour in Emotive Expressions. *Working Papers*, 35, 89–98. Dept of Linguistics. Lund University.

Håkansson, G., & Hansson, K. (2000). Comprehension and production of relative clauses: a comparison between Swedish impaired and unimpaired children. *Journal of Child Language*, 27, 313–333.

Håkansson, G., & Nettelbladt, U. (1996). Similarities between SLI and L2 children. Evidence from the acquisition of Swedish word order. In J. Gilbert, & C. Johnson (Eds.), *Children's language*, vol 9. (pp. 135–151). Mahwah, NJ: Erlbaum.

Hall, E. T. (1966). *The Hidden Dimension.* New York: Doubleday.

Hall, E. T. (1975). Proxemics. In S. Weitz (Ed.), *Nonverbal Communication. Readings with commentary.* (pp. 293–312). New York: Oxford University Press.

Halliday, M. A. K. (1973). *Explorations in the Functions of Language.* London: Edward Arnold.

Halliday, M. (1975). *Learning how to mean. Explorations in the development of language.* London: Edward Arnold.

Hallix, W. A., & Rumbaugh, D. M. (2004). *Animal bodies, human minds: ape, dolphin, and parrot language skills.* New York: Kluwer Academic

Hare, B., Brown, M., Williamson, C., & Tomasello, M. (2002). The domestication of social cognition in dogs. *Science* 298, 1634–1636.

Hare, B., & Tomasello, M. (2004). Chimpanzees are more skillful in competitive than in cooperative cognitive tasks. *Animal Behaviour*, 68, 71–581.

Hare, B., & Tomasello, M. (2005). Human-like social skills in dogs. *Trends in Cognitive Sciences*, 9, 439–444.

Harrington, F. H., & Asa, C. S. (2003). Wolf Communication. In L. D. Mech, & L. Boitani (Eds.), *Wolves. Behaviour, Ecology, and Conservation.* (pp. 66–103). Chicago. The University of Chicago Press.

Harrington, F. H., & Mech, L. D. (1978). Wolf Vocalization. In R. L. Hall, & H. S. Sharp (Eds.), *Wolf and Man: Evolution in Parallel.* (pp. 109–132). New York: Academic Press.

Harrington, F. H., & Mech, L. D. (1983). Wolf Pack Spacing: Howling as a Territory-Independent Spacing Mechanism in a Territorial Population. *Behav Ecol Sociobiol.* 12, 161–168.

Hasselquist, D. (1994). *Male attractiveness, mating tactics and realized fitness in the polygynous Great Reed Warbler.* PhD dissertation. Lund University. Lund, Sweden.

Hasselquist, D. (1998). Polygyny in the great reed warbler: a long term study of factors contributing to male fitness. *Ecology*, 79, 2376–2390.

Hasselquist, D., Bensch, S., & von Schantz, T. (1996). Correlation between male song repertoire, extra-pair paternity and offspring survival in the great reed warbler. *Nature*, 381, 229–232.

Hauk, O., Johnsrude, I., & Pulvermüller, F. (2004). Somatotopic representation of action words in human motor and premotor cortex. *Neuron*, 41, 301–307.

Hauser, M. D. (1996). *The Evolution of communication.* Cambridge, Mass: MIT Press.

Hauser, M., Chomsky, N., & Fitch, W. T. (2002). The faculty of language: what is it, who has it, and how did it evolve? *Science*, 298, 1569–1579.
Hauser, M. D., & Konishi, M. (Eds.). (1999). *The Design of Animal Communication*. Cambridge, Mass.: MIT Press.
Hauser, M. D., & Marler, P. (1993a). Food-associated calls in rhesus macaques (*Macaca mulatta*) I: Socioecological factors influencing call production. *Behavioral Ecology*, 4, 194–205.
Hauser, M. D., & Marler, P. (1993b). Food-associated calls in rhesus macaques (*Macaca mulatta*) II: Costs and benefits of call production and suppression *Behavioral Ecology*, 4, 206–212.
Hayes, K. J. (1950). Vocalizations and speech in chimpanzees. *Amer. Psychol.* 5, 275.
Hayes, K. J., & Hayes, C. (1952). Imitation in a home-raised chimpanzee. *Journal of Comparative Physiology and Psychology*, 45, 450–459.
Haviland, J. B. (2000). Pointing, gesture spaces, and mental maps. In D. McNeill (Ed.), *Language and gesture*. (pp. 13–46). Cambridge: Cambridge University Press.
Haviland, J. B. (2003). How to point in Zinacantán. In S. Kita, (Ed.), *Pointing. Where Language, Culture and Cognition Meet*. (pp. 139–169). Mahwah, New Jersey: Lawrence Erlbaum.
Heinrich, B. (1988). Why do ravens fear their food? *Condor*, 90, 950–952.
Heinze, J., Hölldobler, B., & Gert, A. (1999). Reproductive conflict and division of labor in *Eutetramorium mocquerysi*, a Myrmicione Ant without morphologically distinct female reproductives. *Ethology*, 105, 8, 701–717.
Hendriks, M. C. P., & Vingerhoets, A. J. J. M. (2006). Social messages of crying faces: Their influence on anticipated perception, emotions and behavioural responses. *Cognition & Emotion*, 20, 878–886.
Herman, L. M., & Forestall, P. H. (1985). Reporting presence or absence of named objects by a language-trained dolphin. *Neuroscience and Biobehavioral Reviews*, 9, 667–681.
Herman, L. M., Richards, D. G., & Wolz, J. P. (1984). Comprehension of sentences by bottlenosed dolphins. *Cognition*, 16, 129–219.
Herman, L. M., Abichandi, S. L., Elhajj, A. N., Herman, E. Y. K, Sanchez, J. L., & Pack, A. A. (1999). Dolphins (Tursiops truncates) comprehend the referential character of the human pointing gesture. *Journal of Comparative Psychology*, 113, 347–364.
Hess, E. H. (1972). Pupilometrics. In N. Greenfield, & R. Sternbach (Eds.), *Handbook of Psychophysiology*. New York: Holt, Rinehardt & Winston.
Heymann E. W. (2000). Spatial patterns of scent marking in wild moustached tamarins, Saguinus mystax: no evidence for a territorial function. *Animal Behaviour*, 60, 723–730.
Heymann, E. W. (2006). The Neglected Sense-Olfaction in Primate Behavour, Ecology and Evolution. *American Journal of Primatology*, 68, 519–524.
Hill, P. S. M. (2001). Vibration as a Communication Channel: A Synposis. *American Zoologist*, 41, 1133–1134.
Hockett, C. (1968). The problem of universals in language. In J. Greenberg (Ed.), *Universals of Language*. (pp. 1–29). Cambridge, Mass.: MIT Press.
Hohmann, G. (1989). Vocal Communication of Wild Bonnet Macaques (*Macaca radiata*). *Primates*, 30, 325–345.
Hohmann, G., & Fruth, B. (2003). Culture in Bonobos? Between-species and within-species variation in behavior. *Current Anthropology*, 44, 563–571.
Hollén, L. I., & Radford, A. N. (2009). The development of alarm call behaviour in mammals and birds. *Animal Behaviour*, 78, 791–800.
Hollien, H. (2002). *Forensic Voice Identification*. London: Academic Press.

Hooff, J. v. (1972). A comparative approach to the phylogeny of laughter and smiling. In R. A. Hinde (Ed.), *Non-verbal communication*. (pp. 209-240). Cambridge: Cambridge University Press.

Horowitz, A. (2009). Attention to attention in domestic dog (Canis familiaris) dyadic play. *Animal Cognition*, 12, 107-118.

Houston-Price, C., Plunkett, K., & Harris, P. (2005). 'Word-learning wizardry' at 1;6. *Journal of Child Language*, 32, 175-189.

Hrdy, Blaffer S. (2009). *Mothers and others*. Cambridge, Mass.: Harvard University Press.

Hudson, R. (1999). From molecule to mind: the role of experience in shaping olfactory function. *Journal of Comparative Physiology*, 185, 297-304.

Hughes, M. (1996). The function of concurrent signals: visual and chemical communication in snapping shrimp. *Animal Behaviour*, 52, 247-257.

Hultsch, H., & Todt, D. (2004). Learning to sing. In P. Marler, & H. Slabbekoorn (Eds.), *Nature's music: the science of birdsong*. (pp. 80-107). Amsterdam: Elsevier.

Hunter, M. L., & Krebs, J. R. (1979). Geographical variation in the song of the great tit (Parus major) in relation to ecological factors. *Journal of Animal Ecology*, 48, 759-785.

Hurford, J. R. (1991). The evolution of the critical period for language acquisition. *Cognition*, 40, 159-201.

Hurford, J. R., Studdert-Kennedy, M., & Knight, C. (Eds.). (1998). *Approaches to the Evolution of Language*. Cambridge: Cambridge University Press.

Hurst, J. A., Baraitser, M., Auger, E., Graham, F., & Norell, S. (1990). An extended family with a dominantly inherited speech disorder. *Developmental Medicine and Child Neurology*, 32, 347-355.

Hutchison, L., & Wenzel, B. M. (1980). Olfactory guidance in foraging by procellariiforms. *Condor*, 82, 314-319.

Iversen, P., Kuhl, P. K., Akahane-Yamada, R., Diesch, E., Tohkura, Y., Katterman, A. & Siebert, C. (2003). A perceptual interference account of acquisition difficulties for non-native phonemes. *Cognition*, 87, B47-B57.

Jackendoff, R. (2002). *Foundations of language. Brain, meaning, grammar, evolution*. Oxford: Oxford University Press.

Jacob, J., Balthazart, J., & Schoffeniels, E. (1979). Sex differences in the chemical composition of uropygial gland waxes in domestic ducks. *Biochemical Systematics and Ecology*, 7, 149-153.

Jakobson, R. (1960). Closing statements: Linguistics and poetics. In T. A. Sebeok (Ed.), *Style in Language*. (pp. 350-377). Cambridge, Mass.: MIT Press.

Jang, Y., & Greenfield, M. D. (1996). Ultrasonic communication and sexual selection in wax moths: female choice based on energy and asynchrony of male signals. *Animal Behaviour*, 51, 1095-1106.

Janik, V. M. (2000). Whistle matching in wild bottlenose dolphins (Tursiops truncates). *Science*, 289, 1355-1357.

Jarvis, E. D. (2007). Neural systems for vocal learning in birds and humans: a synopsis. *Journal of Ornithology*, 148, S35-S44.

Jensen, P. (Ed.). (2007). *The behavioural biology of dogs*. Wallingford, Oxfordshire: CABI International Press.

Johansson, S. (2005). *Origins of Language. Constraints on hypothesis*. Amsterdam: John Benjamins.

Johnson, M., Aref, S., & Walters, J. R. (2008). Parent-offspring communication in the western sandpiper. *Behavioural Ecology*, 19, 489-501.

Jönsson, F. U., Olsson, H., & Olsson, J. (2005). Odor emotionality affects the confidence in odor naming. *Chemical Senses*, 30, 29-35.

Joseph, J. E. (2004). *Language and Identity. National, Ethnic, Religious.* New York: Palgrave, Macmillan.

Joslin, P. W. B. (1967). Movements and Home Sites of Timber Wolves in Alonquin Park. *American Zoologist*, 7, 279–288.

Jusczyk, P. W. (1997). *The Discovery of Spoken Language.* Cambridge, Mass.: MIT Press.

Kamil, Alan C. (1998). On the Proper Definition of Cognitive Ethology. In R. P. Balda, I. M. Pepperberg, & A. C. Kamil (Eds.), *Animal Cognition in Nature. The Convergence of Psychology and Biology in Laboratory and Field.* (pp. 1–28). London: Academic Press.

Kaminski, J., Call, J., & Fischer, J. (2004). Word learning in a domestic dog: evidence for fast mapping. *Science*, 304, 1682–1683.

Kano, T. (1982). The Social Group of Pygmy Chimpanzees (Pan paniscus) of Wamba. *Primates*, 23, 171–188.

Kano, T. (1992). *The Last Ape: Pygmy Chimpanzee Behaviour and Ecology.* Stanford: Stanford University Press.

Kappeler, P. M., & Silk, J. B. (Eds.). (2010). *Mind the gap: Tracing the Origins of Human Universals.* Berlin: Springer-Verlag.

Kardong, K. V., & Mackessy, S. P. (1991). The strike behavior of a congenitally blind rattlesnake. *Journal of Herpetology*, 25, 208–211.

Karlsson, P., & Lüscher, M. (1959). Pheromones: a new term for a class of biologically active substances. *Nature*, 183, 55–56.

Katti, M., & Warren, P. S. (2004). Tits, noise and urban bioacoustics. *Trends in Ecology and Evolution*, 19, 109–110.

Kegl, J. (2004). Language Emergence in a Language-Ready Brain: Acquisition Issues. In L. Jenkins (Ed.), *Biolinguistics and the Evolution of Language.* (pp. 207–254). Amsterdam: John Benjamins.

Kellogg, W. N. (1968). Communication and language in a home-raised chimpanzee. *Science*, 182, 423–427.

Kellogg, W. N., & Kellogg, L. A. (1933/1967). *The Ape and the Child. A comparative study of the environmental influence upon early behavior.* New York: Hafner Publishing Company.

Kendon, A. 1975. Movement coordination in social interaction: some examples described. In S. Weitz (Ed.), *Nonverbal Communication. Readings with commentary.* (pp. 119–134). New York: Oxford University Press.

Kendon, A. (1980). Gesticulation and speech: Two aspects of the process of utterance. In: M. R. Key (Ed.), *The relationship of verbal and nonverbal communication.* (pp. 207–227). The Hague: Mouton.

Kendon, A. (2004). *Gesture. Visible action as utterance.* Cambridge. Cambridge University Press.

Kent, R. D., & Miolo, G. (1995). Phonetic abilities in the first year of life. In P. Fletcher & B. McWhinney (Eds.), *The Handbook of Child Language.* (pp. 303–334). Oxford: Blackwell.

Kernan, K. (1969). The acquisition of language by Samoan children. *Working Papers of the Language Behavior Research Laboratory, No 21.* Washington D.C.

Kerswell, K. J., Butler, K. L., Bennett, P., & Hemsworth, P. H. (2010). The relationships between morphological features and social signalling behaviours in juvenile dogs: The effect of early experience with dogs of different morphotypes. *Behavioural Processes*, 85, 1–7.

Kimura, M., & Daibo, I. (2006). Interactional synchrony in conversations about emotional episodes: A measurement by the between-participants pseudosynchrony experimental paradigm. *Journal of Nonverbal Behaviour*, 30, 115–126.

King, B. J. (Ed.). (1999). *The origins of language: What nonhuman primates can tell us.* Santa Fe, N.M.: School of American Research Press.

King, M-C., & Wilson, A. C. (1975). Evolution at two levels in humans and chimpanzees, *Science*, 188, 107–116.

Kipper, S., Mundry, R., Sommer, C., Hultsch, H., & Todt, D. (2006). Song repertoire size is correlated with body measures and arrival date in common nightingales, *Luscinia megarhynchos. Animal Behaviour*, 71, 211–217.

Kita, S. (Ed.). (2003). *Pointing. Where Language, Culture and Cognition Meet*. Mahwah, New Jersey: Lawrence Erlbaum.

Kita, S., & Essegby, J. (2001). Pointing left in Ghana: How a taboo on the use of the left hand influences gestural practice. *Gesture*, 1, 73–95.

Klauer, G., Burda, H., & Nevo, E. (1997). Adaptive differentiations of the skin of the head in a subterranean rodent, Spalax ehrenbergi. *J. Morph.* 233, 53–66.

Klaus, M. H., Kennell, J. H., Plumb, N., & Zuehlbe, S. (1970). Human maternal behaviour at the first contact with her young. *Paediatrics*, 46, 187–192.

Kobayashi, H., & Kohshima, S. (2001). Unique morphology of the human eye and its adaptive meaning: Comparative studies of external morphology of the primate eye. *Journal of Human Evolution*, 40, 419–435.

Kortlandt, A. (1962). Chimpanzees in the wild. *Scientific American*, 206, 128–138.

Kövecses, Z. (2002). *Metaphor. A practical introduction*. Oxford: Oxford University Press.

Krause, J., Lalueza-Fox, C., Orlando, L., Enard, W., Green, R. E., Burbano, H. A., Hublin, J. J., Hänni, C., Fortea, J., de la Rasilla, M., Bertranpetet, J., Rosas, A., & Pääbo, S. (2007). The derived FOXP2 variant of modern humans was shared with Neandertals. *Current Biology*, 17 (21), 1908–1910.

Krauss, R. M., Chen, Y., & Gottesman, R. F. (2000). Lexical gestures and lexical access: A process model. In D. McNeill (Ed.), *Language and gesture*. (pp. 261–283). Cambridge: Cambridge University Press.

Krebs, J. R., & Dawkins, R (1984). Animal signals: mind-reading and manipulation. In J. R. Krebs, & N. B. Davies (Eds.), *Behavioural ecology: an evolutionary approach*. (pp. 380–402). Oxford: Blackwell.

Kroodsma, D. E. (1996). Ecology of passerine song development. In D. E. Kroodsma, & E. H. Miller (Eds.), *Ecology and evolution of acoustic communication in birds*. (pp. 3–19). Ithaca, N.Y.: Cornell University Press.

Kroodsma, D. E. (2004). The diversity and plasticity of birdsong. In P. Marler, & H. Slabbekoorn (Eds.), *Nature's Music: the Science of Birdsong*. (pp. 108–131). San Diego, California: Elsevier Science.

Kroodsma, D. E., & Miller, E. H. (Eds.). (1996). *Ecology and Evolution of Acoustic Communication in Birds*. Ithaca, New York: Cornell University Press.

Kuhl, P. K. (1999). Speech, Language, and the Brain. In M. D. Hauser, & M. Konishi (Eds.), *The Design of Animal Communication*. (pp. 419–450). Cambridge, Mass.: MIT Press.

Kuhl, P. K., Andruski, J. E., Chistovich, I. A., Chistovich, L. A., Kozhevnikova, E. V., Ryskina, V. L., Stolyarova, E. I., Sundberg, U. & Lacerda, F. (1997). Cross-Language Analysis of Phonetic Units in Language Addressed to Infants. *Science*, 277, 684.

Kuhl, P. K., & Miller, J. D. (1975). Speech perception by the chinchilla: Voiced-voiceless distinction in alveolar plosive consonants. *Science*, 190, 69–72.

Kuhl, P. K., & Padden, D. M. (1983). Enhanced discriminability at the phonetic boundaries for the place feature in macaques. *J. Acoust. Soc. Am.*, 73, 3, 1003–1010.

Kulick, D. (1992). *Language shift and cultural reproduction*. Cambridge: Cambridge University Press.

Kummer, H. (1971). *Primate societies: Group Techniques of ecological adaptation*. Chicago: Aldine.

Labov, W. (Ed.). (1972). *Sociolinguistic Patterns.* Philadelphia: University of Philadelphia Press.

Ladygina-Kohts, N. N. (1935/2002). *Infant Chimpanzee and Human Child: A Classic 1935 Comparative Study of Ape Emotions and Intelligence.* F. B. M. de Waal, (Ed.), New York: Oxford University Press.

Lai, C. S. L., Fisher, S. E., Hurst, J. A., Vargha-Khadem, F., & Monaco, A. P. (2001). A forkhead-domain gene is mutated in a severe speech and language disorder. *Nature,* 413, 519–523.

Laidre, M. (2012). Mandrill visual gestures. A round-the-world study of the largest of all monkeys. In S. Pika, & K. Liebal (Eds.), *Developments in Primate Gesture Research.* (pp. 113–128). Amsterdam: John Benjamins.

Lakin, J. L., Jefferis, V. E., Cheng, C. M., & Chartrand, T. L. (2003). The chameleon effect as social glue: Evidence for the evolutionary significance of nonconscious mimicry. *Journal of Nonverbal Behavior,* 27, 3, 145–161.

Lakoff, G., & Johnson, M. (1980). *Metaphors we live by.* Chicago: University of Chicago Press.

Lamb, S. M. (1999). *Pathways of the Brain. The Neurocognitive Basis of Language.* Amsterdam: John Benjamins.

Lambrechts, M. M. (1992). Male quality and playback in the great tit. In P. K McGregor (Ed.), *Playback and Studies of Animal Communication.* (pp. 135–152). New York: Plenum Press.

Langbauer, W. R. Jr. (2000). Elephant Communication. *Zoo Biology,* 19, 425–445.

Lauay, C., Gerlack, N. M., Adkins-Regan, E., & Devoogd, T. J. (2004). Female zebra finches require early song exposure to prefer high-quality song as adults. *Animal Behaviour,* 68, 1249–1255.

LaVelli, M., & Fogel, A. (2002). Developmental Changes in Mother-Infant Face-to-Face Communication: Birth to 3 Months. *Developmental Psychology,* 38, 288–305.

Lazaro-Perea, C., De Fátima Arruda, M., & Snowdon, C. T. (2004). Grooming as a reward? Social function of grooming between females in cooperatively breeding marmosets. *Animal Behaviour,* 67, 627–636.

Leadbeater, E., Goller, F., & Riebel, K. (2005). Unusual phonation, covarying song characteristics and song preferences in female zebra finches. *Animal Behaviour,* 70, 909–919.

Leavens, D. A., & Hopkins, W. D. (1999). The Whole-Hand Point: The Structure and function of Pointing from a Comparative Perspective. *Journal of Comparative Psychology,* 113, 417–425.

Leavens, D. A., Hopkins, W. D., & Bard, K. A. (1996). Indexical and Referential Pointing in Chimpanzee (*Pan troglodytes*). *Journal of Comparative Psychology,* 110, 346–353.

Leavens, D. A., Hopkins, W. D., & Bard, K. A. (2005). Understanding the Point of Chimpanzee Pointing. *Current Directions in Psychological Science,* 14, 185–189.

Leaver, S. D. A., & Reimchen, T. E. (2008). Behavioural responses of *Canis familiaris* to different tail lengths of a remotely-controlled life-size dog replica. *Behaviour,* 145, 377–390.

Lemasson, A. (2011). What can forest guenons "tell" us about the origin of language? In A. Vilain, J-L. Schwartz, C. Abry, & J. Vauclair (Eds.), *Primate Communication and Human Language. Vocalization, gestures, imitation and deixis in humans and non-humans.* (pp. 39–70). Amsterdam: John Benjamins.

Lemasson, A., Gautier, J-P., & Hausberger, M. (2003). Vocal similarities and social bonds in Campbell's monkey (*Cercopithecus campbelli*). *Current Review of Biologies,* 326, 1185–1193.

Lemasson, A., & Hausberger, M. (2004). Patterns of vocal sharing and social dynamics in a captive group of Campbell's monkeys. *Journal of Comparative Psychology,* 118, 347–359.

Lenneberg, E. H. (1967). *Foundations of language.* New York: Wiley.

Leonard, L. B. (1998). *Children with Specific Language Impairment.* Cambridge, Mass.: MIT Press.

Levelt, W. J. M. (1989). *Speaking: from intention to articulation.* Cambridge Mass.: MIT Press.

Levinson, S. C., & Wilkins, D. (2006). Patterns in the data: towards a semantic typology of spatial description. In S. C. Levinson, & D. Wilkins (Eds.), *Grammars of Space. Explorations in Cognitive Diversity*. (pp. 512–552). Cambridge: Cambridge University Press.

Lewis, E. R., & Narins, P. M. (1985). Do frogs communicate with seismic signals? *Science*, 227, 187–189.

Liddell, S. K. (2003). *Grammar, gesture and meaning in American Sign Language*. Cambridge: Cambridge University Press.

Lieberman, P. (1991). *Uniquely human. The evolution of speech, thought, and selfless behavior*. Cambridge, Mass.: Harvard University Press.

Lieberman, P. (2000). *Human language and our reptilian brain - the subcortical bases of speech, syntax, and thought*. Cambridge Mass.: Harvard University Press.

Lieberman, P. (2006). The FOXP2 gene, human cognition and language. *International Congress Series*, 1296, 115–126.

Linell, P. (2009). *Rethinking Language, Mind and World Dialogically: Interactional and contextual theories of human sense-making*. Charlotte, NC: Information Age Publishing.

Linnaeus, C. (1758). Systema naturae per regna tria naturae, secundum classes, ordines, genera, species, cum characteribus, differentiis, synonymis, locis. Editio decima, reformata. Holmiae. (Laurentii Salvii).: [1–4], 1–824.

Linville, S. E. (2001). *Vocal Ageing*. San Diego: Singular. Thomson Learning.

Liszkowski, U., Carpenter, M., Henning, A., Striano, T., & Tomasello, M. (2007). Twelve-months-olds point to share attention and interest. *Developmental Science*, 7 (3), 297–307.

Liu, W.-C., & Kroodsma, D. E. (2006). Song learning by chipping sparrows: when, where, and from whom. *The Condor*, 108, 509–517.

Lombard, E. (1911). Le signe de l'elevation de la voix. *Annales de Maladies de L'oreille et du Larynx*, 37, 101–119.

Lorenz, K. (1935). Der Kumpan in der Umwelt des Vogels. *Journal of Ornithology*, 83, 137–413.

Lorenz, K. (1943). Die angeborden Formen möglicher Erfahrung. *Zeitung f Tierpsychol*, 5, 235–409.

Lorenz, K. (1954). *Man meets dog*. London: Methuen.

Lorenz, K. (1981). *The Foundations of Ethology*. New York: Simon and Schuster.

Losey, G. S. Jr. (2003). Crypsis and communication functions of UV-visible coloration in two coral reef damselfish, *Dascyllus aruanus* and *D. reticulates*. *Animal Behaviour*, 66, 299–307.

Lusseau, D., Wilson, B., Hammond, P. S., Grellier, K., Durban, J. W., Parsons, K. M., Barton, T. R., & Thompson, P. M. (2006). Quantifying the influence of sociality on population structure in bottlenose dolphins. *Journal of Animal Ecology*, 75, 14–24.

Lyn, H. (2012). Apes and the Evolution of Language: Taking Stock of 40 Years of Research. In J. Vonk, & T. K. Shackelford (Eds.), *The Oxford Handbook of Comparative Evolutionary Psychology*. (pp. 356–378). Oxford: Oxford University Press.

Mack, A. L., & Jones, J. (2003). Low-frequency vocalizations by Cassowaries (*Casuarius Spp.*) *The Auk*, 120, 1062–1068.

MacWhinney, B. (2000). *The CHILDES Project: Tools for Analyzing Talk. Volume 1: Transcription format and programs. Volume 2: The Database*. Mahwah, NJ: Lawrence Erlbaum Associates. (These are links to recent electronic revisions).

Maestripieri, D., Ross, S. K., & Megna, N. L. (2002). Mother-infant interaction in western lowland gorillas (Gorilla gorilla gorilla): spatial relationships, communication, and opportunities for social learning. *Journal of Comparative Psychology*, 116, 219–227.

Maklakov, A. A., Bilde, T., & Lubin, Y. (2003). Vibratory courtship in a web-building spider: signalling quality or stimulating the female? *Animal Behaviour*, 66, 623–630.

Makwana, S. C. (1978). Field Ecology and Behaviour of the Rhesus Macaque (Macaca mulatta):. Group Composition, Home Range, Roosting Sites, and Foraging Routes in the Asarori Forest. *Primates*, 19, 483–492.

Mann, N. J., & Slater, P. J. B. (1994). What causes young male zebra finches, *Taeniopygia guttata*, to choose their father as song tutor? *Animal Behaviour*, 47, 671–677.

Mann, N. J., & Slater, P. J. B. (1995). Song tutor choice by zebra finches in aviaries. *Animal Behaviour*, 49, 811–820.

Manning, A., & Dawkins, M. S. (1992). *An introduction to animal behaviour*. Cambridge: Cambridge University Press.

Månsson, A-C. (2003). *The relation between gestures and semantic processes. A study of normal language development and specific language impairment in children*. Diss. Dept of Linguistics. Göteborg University.

Marcus, G. F. (2006). Startling starlings. *Nature*, 440, 1117–1118.

Markson, L., & Bloom, P. (2001). Perceiving Intentions and Learning Words in the Second Year of Life. In M. Tomasello, & E. Bates (Eds.), *Language Development. The Essential Readings*. (pp. 129–134). Oxford: Blackwell.

Marler, P. (1952). Variation in the song of the chaffinch, *Fringilla coelebs*. *Ibis*, 94, 458–472.

Marler, P. (1970a). A comparative approach of vocal learning: song development in White-Crowned Sparrows. *Journal of Comparative Physiology and Psychology*, 71, (suppl) 1–25.

Marler, P. (1970b). Birdsong and human speech: Could there be parallels? *American Scientist*, 58, 669–674.

Marler, P. (1999). On Innateness: Are Sparrow Songs "Learned" or "Innate"? In M. D. Hauser, & M. Konishi (Eds.), *The Design of Animal Communication*. (pp. 293–318). Cambridge, Mass.: MIT Press.

Marler, P. (2004a). Science and birdsong: the good old days. In P. Marler, & H. Slabbekoorn (Eds.), *Nature's music: the science of birdsong*. (pp. 1–38). Amsterdam: Elsevier.

Marler, P. (2004b). Bird calls: a cornucopia for communication. In P. Marler, & H. Slabbekoorn (Eds.), *Nature's music: the science of birdsong*. (pp. 132–177). Amsterdam: Elsevier.

Marler, P., Dufty, A., & Pickert, R. (1986a). Vocal communication in the domestic chicken. I. Does a sender communicate information about the quality of a food referent to a receiver? *Animal Behaviour*, 34, 188–193.

Marler, P., Dufty, A., & Pickert, R. (1986b). Vocal communication in the domestic chicken. II. Is a sender sensitive to the presence and nature of a receiver? *Animal Behaviour*, 34, 194–198.

Marler, P., Evans, C. S., & Hauser, M. D. (1992). Animal signals: Motivational, referential, or both? In H. Papousek, U. Jurgens & M. Papousek (Eds.), *Nonverbal Vocal communication*. (pp. 66–86). Cambridge: Cambridge University Press.

Marler, P., Evans, C. S., & Evans, L. (1993). On the meaning of alarm calls: functional reference in an avian vocal system. *Animal Behaviour*, 46, 23–38.

Marler, P., & Peters, S. (1982). Developmental overproduction and selective attrition: new processes in the epigenesis of birdsong. *Devel. Psychobiol.* 15, 369–378.

Marler, P., & Slabbekoorn, H. (Eds.). (2004). *Nature's music: the science of birdsong*. Amsterdam: Elsevier.

Marsh, A. A., Adams, R. B., & Kleck, R. E. (2005). Why do fear and anger look the way they do? Form and social function in facial expressions. *Personality and Social Psychology Bulletin*, 31, 73–86.

Marshall-Ball, L., Mann, N., & Slater, P. J. B. (2006). Multiple functions to duet singing: hidden conflicts and apparent cooperation. *Animal Behaviour*, 71, 823–831.

Martin, G. R. (1986). Total panoramic vision in the mallard duck, Anas Platyrhynchos. *Vision Research*, 26, 8, 1303–1305.

Martin, J., Barja, I., & López, P. (2010). Chemical scent constituents in feces of wild Iberian wolves (*Canin lupus signatus*). *Biochemical Systematics and Ecology*, 38, 1096–1102.

Mathew, R. (1997). English in India: a response to Mark Tully. *ELT Journal*, 51, 165–168.

Mayberry, R. & Jaques, J. (2000). Gesture production during stuttered speech: insights into the nature of gesture – speech integration. In D. McNeill (Ed.), *Language and gesture*. (pp. 199–214). Cambridge: Cambridge University Press.

Mayberry, R., & Squires, B. (2006). Sign Language Acquisition. In E. Lieven (Ed.), *Language Acquisition. Encyclopedia of Language and Linguistics*. (pp. 291–296). Oxford: Elsevier.

Maynard Smith, J., & Harper, D. (2003). *Animal signals*. Oxford: Oxford University Press.

McClintock, M. K. (1971). Menstrual synchrony and suppression. *Nature*, 229, 244–245.

McConnell, P. B. (1991). Lessons from animal trainers: the effect of acoustic structure on an animal's response. In P. Bateson, & P. Klopfer (Eds.), *Perspectives in Ethology*. (pp. 165–187). New York: Plenum Press.

McConnell, P. B., & Baylis, J. R. (1985). Interspecific communication in cooperative herding: acoustoc and visual signals from human shepherds and herding dogs. *Z. Tierpsychologie*, 67, 302–328.

McCowan, B., Hanser, S. F., & Doyle, L. R. (1999). Quantitative tools for comparing animal communication systems: information theory applied to bottlenose dolphin whistle repertoires. *Animal Behaviour*, 57, 409–419.

McCowan, B., & Reiss, D. (2001). The fallacy of 'signature whistles' in bottlenose dolphins; a comparative perspective of 'signature information' in animal vocalizations. *Animal Behaviour*, 62, 1151–1162.

McCune, L., Vihman, M. M., Rough-Hellichius, L., Delery, D. B., & Gogate, L. (1996). Grunt communication in human infants (Homo sapiens). *Journal of Comparative Psychology*, 110, 27–36.

McGregor, P. (Ed.). (2005). *Animal Communication Networks*. Cambridge: Cambridge University Press.

McGregor, P., & Peake, T. M. (2000). Communication networks: social environments for receiving and signalling behaviour. *Acta Ethologica*, 2, 71–81.

McGregor, P. K., Dabelsteen, T., Shepherd, M., & Pedersen, S. B. (1992). The signal value of matched singing in great tits: evidence from interactive playback experiments. *Animal Behaviour*, 43, 987–998.

McGrew, W. (2004). *The Cultured Chimpanzee. Reflections on Cultural Primatology*. Cambridge: Cambridge University Press.

McKinley, J., & Sambrook, T. D. (2000). Use of human-given cues by domestic dogs (Canis familiaris) and horses (Equus caballus). *Animal Cognition*, 3, 13–22.

McNeill, D. (1985). So you think gestures are nonverbal? *Psychological Review*, 92, 350–371.

McNeill, D. (1992). *Hand and mind. What gestures reveal about thought*. Chicago: University of Chicago Press.

McNeill, D. (2005). *Gesture & Thought*. Chicago: University of Chicago Press.

Mech, L. D. (1970). *The Wolf. The Ecology and Behavior of an Endangered Species*. New York: Doubleday Publishing.

Mech, L. D. (1966). *The wolves of Isle Royale*. U.S. National Park Service Fauna Series no 7. Washington D.C.: Government Printing Office.

Mech, L. D., & Boitani, L. (Eds.). (2003). *Wolves. Behavior, Ecology, and Conservation*. Chicago: University of Chicago Press.

Medjo, D. C., & Mech, L. D. (1976). Reproductive activity in nine and ten month old wolves. *J. Mamm.*, 57, 406–408.

Meisel, J. (2011). *First and second language acquisition*. Cambridge: Cambridge University Press.

Meltzoff, A. N. (1981). Imitation, intermodal coordination, and representation in early infancy. In G. Butterworth (Ed.), *Infancy and epistemology*. (pp. 85–113). Brighton, England: Harvester Press.

Meltzoff, A. N., & Moore, M. K. (1977). Imitation of facial and manual gestures by human neonates. *Science*, 198, 75–78.

Menzel, C. R. (1999). Unprompted recall and reporting of hidden objects by a chimpanzee (Pan troglodytes) after extended delays. *Journal of Comparative Psychology*, 113, 426–434.

Merker, B., & Cox, C. (1999). Development of the female great call in Hyltobates gabriellae: A case study. *Folia Primatologica*, 70, 97–106.

Meunier, H., Deneubourg, J-L., & Petit, O. (2008). How many for dinner? Recruitment and monitoring by glances in capuchins. *Primates*, 49, 26–31.

Michelsen, A. (1999). The Dance Language of the Honeybees: Recent findings and Problems.. In M. D. Hauser, & M. Konishi (Eds.), *The Design of Animal Communication*. (pp. 111–131). Cambridge, Mass.: MIT Press.

Mignault, A., & Chadhuri, A. (2003). The many faces of a neutral face: head tilt and perception of dominance and emotion *Journal of Nonverbal Behavior*, 27, 111–132.

Miklósi, Á. (2007a). *Dog behaviour, evolution and cognition*. Oxford: Oxford University Press.

Miklósi, Á. (2007b). Human-Animal Interactions and Social Cognition in Dogs. In P. Jensen (Ed.), *The behavioural biology of dogs*. (pp. 207–222). Wallingford, Oxfordshire: CABI International Press.

Miklósi, Á., Polgárdi, R., Topál, J., & Csányi, V. (2000). Intentional behaviour in dog-human communication: An experimental analysis of "showing" behaviour in the dog. *Animal Cognition*, 3, 159–166.

Miklósi, Á., Kubinyi, E., Tópal, J., Gáesi, M., Virányi, Z., & Csányi, V. (2003). A simple reason for a big difference: wolves do not look back at humans but dogs do. *Current Biology*, 13, 763–767.

Miklósi, Á., & Soprani, K. (2006). A comparative analysis of animals' understanding of the human pointing gesture. *Animal Cognition*, 9, 81–93.

Miles, H. L. (1983). Apes and Language: The Search for Communicative competence. In J. de Luce, & H. T. Wilder (Eds.), *Language in Primates*. (pp. 43–61). New York: Springer Verlag.

Miller, H. C., Rayburn-Reeves, R., & Zentall, T. R. (2009). Imitation and emulation by dogs using a bidirectional control procedure. *Behavioural Processes*, 80, 109–114.

Miller, K., Laszlo, K., & Dietz, J. M. (2003). The role of scent marking in the social communication of wild golden lion tamarins, *Leontopitheus rosalia*. *Animal Behaviour*, 65, 795–803.

Millikan, R. G. (2004). On Reading Signs: Some Differences Between Us and the others. In D. Oller, & U. Griebel (Eds.), *The Evolution of Communication Systems: A Comparative Approach*. (pp. 15–29). The Vienna Series in Theoretical Biology, Cambridge Mass.: MIT Press.

Mills, D. S. & McDonnell, S. M. (2005). *The domestic horse: the origins, development, and management of its behaviour*. Cambridge: Cambridge University Press.

Milne, A. A. (1965). *Winnie-the-Pooh*. London: Methuen & Co.

Milne, A. A. (1979). *Nalle Puh*. (Swedish translation). Stockholm: Bonniers.

Mitani, J. C. (1994). Ethological studies of chimpanzee vocal behavior. In R. Wrangham, W. McGrew, F. DeWaal, & P. Heltne (Eds.), *Chimpanzee Cultures*. (pp. 195–210). Cambridge: Harvard University Press.

Mitani, J. C., & Amsler, S. J. (2003). Social and spatial aspects of male subgrouping in a community of wild chimpanzees. *Behaviour*, 140, 869–884.

Mitani, J. C., & Gros-Louis, J. (1998). Chorusing and call convergence in chimpanzees: test of three hypotheses. *Behaviour*, 135, 1041–1064.

Mitani, J. C., Hasegawa, T., Gros-Louis, J., Marler, P., & Byrne, R. (1992). Dialects in wild chimpanzees? *American Journal of Primatology*, 27, 233–252.

Mithen, S. (2005). *The singing Neanderthals. The origins of music, language, mind and body.* London: Weidenfeld & Nicolson.

Møller, A. P. (1990). Male tail length and female mate choice in the monogamous swallow *Hirando rustica*. *Animal Behaviour*, 39, 458–465.

Moncomble, A-S., Coureaud, G., Quennedey, B., Langlois, D., Perrier, G., & Schaal, B. (2005). The mammary pheromone of the rabbit: from where does it come? *Animal Behaviour*, 69, 29–38.

Morris, D., Collett, P., Marsh, P., & O'Shaugnessy, M. (1979). *Gestures. Their origins and distribution.* London: Jonathan Cape.

Morton, E. S. (1977). On the occurrence and significance of motivation-structural rules in some bird and mammal sounds. *American Nature*, 111, 855–869.

Morton, E. S. (1982). Grading, Discreteness, Redundancy, and Motivation-Structural Rules. In D. E. Kroodsma, & E. H. Miller (Eds.), *Acoustic Communication in Birds*, Vol 1. (pp. 183–212). New York: Academic Press.

Morton, E. S. (1996). A comparison of vocal behavior among tropical and temperate passerine birds. In D. E. Kroodsma, & E. H. Miller (Eds.), *Ecology and Evolution of Acoustic Communication in Birds.* (pp. 258–268). Ithaca, New York: Cornell University Press.

Mowat, L. (1965). *Never Cry Wolf.* New York: Dell Publ.

Murie, A. (1944). *The wolves of Mount McKinley.* Washington D.C.: National Park Service Fauna Series, no 5. U.S. Government Printing Office.

Naderi, Sz., Miklósi, A., Dóka, A., & Csányi, V. (2001). Co-operative interactions between blind persons and their dogs. *Applied Animal Behaviour Science*, 74, 59–80.

Naguib, M. (1996). Auditory distance estimation in song birds: Implications, methodologies and perspectives. *Behavioural Processes*, 38, 163–168.

Naguib, M. & Wiley, R. H. (2001). Estimating the distance to a source of sound: mechanisms and adaptations for long-range communication. *Animal Behaviour*, 62, 825–837.

Nakamura M. (2002). Grooming-hand-clasp in Mahale M group chimpanzees: Implications for culture in social behaviors. In C. Boesch, G. Hohmann, & L. F. Marchant (Eds.), *Behavioural Diversity in Chimpanzees and Bonobos.* (pp. 71–83). Cambridge: Cambridge University Press.

Nakamura, M., & Itoh, N. (2005). Notes on the behavior of a newly immigrated female chimpanzee to the Mahale M group. *Pan Africa News*, 12, 20–22.

Narins, P. M., Lewis, E. R., Jarvis, J. J. U. M., & O'Riain, J. (1997). The use of seismic signals by fossorial southern African mammals: a neuroethological gold mine. *Brain Research Bulletin*, 44(5), 641–646.

Neill, S. R. St. J. (1986). Children's reported response to teachers nonverbal signals: a pilot study. *Journal of Education and Teaching*, 12, 53–63.

Nelson, B. S. (2000). Avian dependence on sound pressure level as an auditory distance cue. *Animal Behaviour*, 59, 57–67.

Nelson, J. K. (2005). *Seeing through tears: crying and attachment.* New York: Brunner-Routledge.

Nelson, D. A., & Soha, J. A. (2004). Perception of geographical variation in song by male Puget Sound white-crowned sparrows, *Zonotricia leucophrys pugetensis*. *Animal Behaviour*, 68, 395–405.

Nevitt, G. A (2000). Olfactory foraging by Antarctic procellariiform seabirds: life at high Reynolds numbers. *Biological Bulletin*, 198, 245–253.

Newbury, D. F., Bonora, E., Lamb, J. A., Fisher, S. E., Lai, C. S. L., Baird, G., Jannoun, L., Slonims, V., Stott, C. M., Merricks, M. J., Bolton, P. F., Bailey, A. J., Monaco, A. P. & the International Molecular Genetic Study of Autism Consortium. (2002). FOXP2 Is Not a Major Susceptibility Gene for Autism or Specific Language impairment. *Am. J. Hum. Genet.* 70, 1318–1327.

Newport, E. L., Gleitman, H., & Gleitman, L. R. 1977. Mother, I'd rather do it myself: some effects and non-effects of maternal speech style. In C. Snow, & C. Ferguson (Eds.), *Talking to children. Language Input & Acquisition.* (pp. 109–149). Cambridge: Cambridge University Press.

Nicoladis, E. (2002). Some gestures develop in conjunction with spoken language development and others don't: evidence from bilingual preschoolers. *Journal of Nonverbal Behavior*, 26, 241–266.

Nishida, T. (1968). The social group of wild chimpanzees in the Mahali Mountains. *Primates*, 9, 167–224.

Nishida, T., Mitani, J. C., & Watts, D. P. (2004). Variable grooming behaviours in wild chimpanzees. *Folia Primatol.* 75, 31–36.

Nishimura, T., Mikami, A., Suzuki, J., & Matsuzawa, T. (2003). Descent of the larynx in chimpanzee infants. *PNAS*, 100, 6930–6933.

Noad, M. J., Cato, D. H., Bryden, M. M., Jenner, M. N., & Jenner, K. C. S. (2000). Culture revolution in whale songs. *Nature*, 408, 537.

Nolan, P. M., & Hill, G. E. (2004). Female choice for song characteristics in the house finch. *Animal Behaviour*, 67, 403–410.

Nord, C. (1997). Alice abroad. Dealing with descriptions and transcriptions of paralanguage in literary translation. In F. Poyatos (Ed.), *Nonverbal Communication and translation.* (pp. 108–129). Amsterdam: John Benjamins.

Nottebohm, F. (1999). The Anatomy and Timing of Vocal Learning in Birds. In M. D. Hauser, & M. Konishi (Eds.), *The Design of Animal Communication.* (pp. 63–110). Cambridge, Mass.: MIT Press.

Nowicki, S., Hasselquist, D., Bensch, S., & Peters, S. (2000). Nestling growth and song repertoire size in great reed warblers: evidence for song learning as an indicator mechanism in mate choice. *Proceedings the Royal Society London*, 267, 2419–2424.

Núnes, R. E., & Sweetser, E. (2006). With the future behind them: Convergent evidence from Aymara language and gesture in the crosslinguistic comparison of spatial construal of time. *Cognitive Science*, 30, 401–450.

Ochs, E. (1988). *Culture and Language Development: language acquisition and language socialization in a Samoan village.* Cambridge: Cambridge University Press.

Ochs, E., & Schieffelin, B. (2001). Language Acquisition and Socialization: Three Developmental Stories and their Implications. In A. Duranti (Ed.), *Linguistic Anthropology. A Reader.* (pp. 263–301). Cambridge: Cambridge University Press.

O'Connor, K. I., Melcalfe, N. B., & Taylor, A. C. (1999). Does darkening signal submission in territorial contests between juvenile Atlantic salmon, *Salmo salar*? *Animal Behaviour*, 58, 1269–1276.

Ohala, J. J. (1980). The acoustic origin of the smile. *Journal of the Acoustical Society of America*, 68, S33–S33.

Ohala, J. J. (1984). An ethological perspective on common cross-language utilization of F0 of voice. *Phonetica*, 41, 1–16.

Oller, D. K. (2004). Underpinnings for a Theory of Communicative Evolution. In D. K. Oller, & U. Griebel (Eds.), *The Evolution of Communication Systems: A Comparative Approach.* (pp. 49–65). The Vienna Series in Theoretical Biology, Cambridge Mass.: MIT Press.

Oller, D. K., Eilers, R. E., Neal, A. R., & Schwartz, H. K. (1999). Precursors to speech in infancy: the prediction of speech and language disorders. *Journal of Communication Disorders*, 32, 223–245.

Oller, D., & Griebel, U. (Eds.). (2004). *The Evolution of Communication Systems: A Comparative Approach*. The Vienna Series in Theoretical Biology, Cambridge Mass.: MIT Press.

Owren, M, J., & Bachorowski, J.-A. (2003). Reconsidering the evolution of non-linguistic communication: the case of laughter. *Journal of Nonverbal Behavior*, 27, 183–200.

Pal, S. K. (2005). Parental care in free-ranging dogs, *Canis familiaris*. *Applied Animal Behaviour Science*, 90, 31–47.

Pal, S. K. (2008). Maturation and development of social behaviour during early ontogeny in free-ranging dog puppies in West Bengal, India. *Applied Animal Behaviour Science*, 111, 95–107.

Palacios, V., Font, E., & Márquez, R. (2007). Iberian wolf howls: acoustic structure, individual variation, and a comparison with North American populations. *Journal of Mammalogy*, 88, 606–613.

Pallier, C., Dehaene, S., Poline, J.-B., LeBihan, D., Argenti, A.-M., Dupoux, E., & Mehler, J. (2003). Brain imaging of language plasticity in adopted adults: can a second language replace the first? *Cerebral Cortex*, 13, 155–161.

Palmer, G., & Brown, R. (1998). The ideology of honour, respect, and emotion in Tagalog. In A. Athanasiadou, & E. Tabakowska (Eds.), *Speaking of emotions: Conceptualisation and expression*. (pp. 331–355). Berlin. Germany: Mouton de Gruyter.

Papke, M. D., Reichert, S. E., & Schulz, S. (2001). An airborne female pheromone associated with male attraction and courtship in a desert spider. *Animal Behaviour*, 61, 877–886.

Papousek, H., & Papousek, M. 1977. Mothering and the Cognitive Head-Start: Psychobiological Considerations. In H. R. Schaffer (Ed.), *Studies in the Mother-Infant Interaction*. (pp. 63–85). London: Academic Press.

Paradis, J., & Crago, M. (2000). Tense and temporality: a comparison between children learning a second language and children with SLI. *Journal of Speech, Language and Hearing Research*, 43, 834–847.

Parisi, D. & Cangelosi, A. (2002). A unified simulation scenario for language development, evolution, and historical change. In A. Cangelosi & D. Parisi (Eds.), *Simulating the evolution of language*. (pp. 255–276). London: Springer.

Parr, L., & Waal, F. de. (1999). Visual recognition in chimpanzees. *Nature*, 399, 647.

Patterson, F. G., & Linden, E. (1981). *The education of Koko*. New York: Holt, Rinehart and Winston.

Patterson, F. G., & Cohn, R. H. (1990). Language acquisition by a lowland gorilla: Koko's first ten years of vocabulary development. *Word*, 41, 97–142.

Paulsell, S., & Goldman, M. (1984). The effect of touching different body areas on prosocial behavior. *Journal of Social Psychology*, 122, 269–273.

Pause, B. M. (2004). Are androgen steroids acting as pheromones in humans? *Physiology & Behavior*, 83, 21–29

Payne, R. (Ed.). (1983). *Communication & Behaviour of Whales*. Boulder. Colorado: Westview Press.

Payne, R., & McVay, S. (1971). Songs of humpback whales. *Science*, 173, 585–597.

Payne, R. B., & Payne, K. (1977). Social organisation and mating success in local song populations of village indigobirds, Vidua chalybeata. *Z. Tierpsychologie*, 45, 113–173.

Peake, T. M., Matessi, G. McGregor, P. K., & Dabelsteen, T. (2005). Song type matching, song type switching and eavesdropping in male great tits. *Animal Behaviour*, 69, 1063–1068.

Pepperberg, I. M. (1998). The African Grey Parrot: How cognitive processing might affect allospecific vocal learning. In R. P. Balda, I. M. Pepperberg, & A. C. Kamil (Eds.), *Animal Cognition in Nature. The Convergence of Psychology and Biology in Laboratory and Field*. (pp. 381–409). London: Academic Press.

Pepperberg, I. M. (1999). *The Alex studies.* Cambridge. Mass.: Harvard University Press,

Pepperberg, I. M. (2004). Grey parrots: learning and using speech. In P. Marler, & H. Slabbekoorn (Eds.), *Nature's music: the science of birdsong.* (pp. 363-373). Amsterdam: Elsevier.

Pepperberg, I. M., & Schinke-Llano, L. (1991). Language Acquisition and Form in a Bilingual Environment: A Framework for Studying Birdsong in Zones of Sympatry. *Ethology,* 89, 1-28.

Peters, R. P., & Mech, L. D. (1975). Scent-marking in wolves. *American Scientist,* 63, 628-637.

Peterson, R. O., & Ciussi, P. (2003). The Wolf as a Carnivore. In L. D. Mech, & L. Boitani (Eds.), *Wolves. Behavior, Ecology, and Conservation.* (pp. 115-130). Chicago: University of Chicago Press.

Petitto, L. A., & Marentette, P. F. (1991). Babbling in the Manual Mode: Evidence for the Ontogeny of Language. *Science,* 251, 1493-1496.

Petrie, M., & Halliday, T. (1994). Experimental and natural changes in the peacock's (*Pavo cristatus*) train can affect mating success. *Behavioral Ecology and Sociobiology,* 35, 213-217.

Piaget, J. (1959). *The Language and Thought of the child.* London: Routledge & Kegan Paul.

Pickering, M. J., & Garrod, S. (2004). Toward a mechanistic psychology of dialogue. *Behavioral and Brain Sciences,* 27, 169-226.

Pickering, M. J., & Garrod, S. (2006). Alignment as the basis for successful communication. *Research on Language and Computation,* 4, 203-228.

Pienemann, M. (1998). *Language processing and second language development. Processability Theory.* Amsterdam: John Benjamins.

Pienemann, M. (Ed.). (2005). *Cross-linguistic aspects of Processability Theory.* Amsterdam: John Benjamins.

Pika, S., Liebal, K., & Tomasello, M. (2003). Gestural Communication in young gorillas (*Gorilla gorilla*): Gestural repertoire, learning and use. *American Journal of Primatology,* 60, 95-111.

Pika, S., Liebal, K., & Tomasello, M. (2005). Gestural Communication in Subadult Bonobos (*Pan paniscus*): Repertoire and Use. *American Journal of Primatology,* 65, 39-61.

Pika, S., & Liebal, K. (2012). *Developments in Primate Gesture Research.* Amsterdam: John Benjamins.

Pinker, S. (1994). *The language instinct. How the mind creates language.* New York: Harper Perennial.

Pinker, S., & Bloom, P. (1990). Natural language and natural selection. *Behavioural and Brain Sciences,* 13, 707-784.

Pistorio, A. L., Vintch, B., & Wang, X. (2006). Acoustic analysis of vocal development in a New World primate, the common marmoset (*Callithrix jacchus*). *Journal of the Acoustic Society in America,* 120(3), 1655-1670.

Plooij, F. (1979). How wild chimpanzee babies trigger the onset of mother-infant play – and what the mother makes of it. In M. Bullowa (Ed.), *Before Speech. The beginning of interpersonal communication* (pp. 223-244). Cambridge: Cambridge University Press.

Podos, J., Sherer, J. K., Peters, S., & Nowicki, S. (1995). Ontogeny of vocal tract movements during song production in song sparrows. *Animal Behaviour,* 50, 1287-1296.

Poesel, A., Nelson, D. A., & Gibbs, H. L. (2012). Song sharing correlates with social but not extrapair mating success in the white-crowned sparrow. *Behavioral Ecology,* 23, 627-634.

Pola, Y. V., & Snowdon, C. T. (1975). The vocalizations of pygmy marmosets (*Cebuella pygmaea*). *Animal Behaviour,* 23, 826-842.

Pongrácz, P., Molnár, C., Miklósi, Á., & Csányi, V. (2005). Human listeners are able to classify dog (*Canis familiaris*) barks recoded in different situations. *Journal of Comparative Psychology,* 119(2), 136-144.

Pongrácz, P., Molnár, C., & Miklósi, A. (2010). Barking in family dogs: An ethological approach. *The Veterinary Journal,* 183, 141-147.

Poole, J. H. (1999). Signals and assessment in African elephants: evidence from playback experiments. *Animal Behaviour*, 58, 185–193.
Poulter, T. C. (1968). Marine mammals. In T. A. Sebeok (Ed.), *Animal Communication. Techniques of Study and Results of Research*. (pp. 405–465). Bloomington: Indiana University Press.
Power, C. (1998). Old wives' tales: the gossip hypothesis and the reliability of cheap signals. In J. R. Hurford, M. Studdert-Kennedy, & C. Knight (Eds.), *Approaches to the Evolution of Language*. (pp. 111–129). Cambridge: Cambridge University Press.
Poyatos, F. (Ed.). (1997). *Nonverbal Communication and translation*. Amsterdam: John Benjamins.
Premack, D. (1971). Language in the chimpanzee? *Science*, 172, 808–822.
Premack, A. J., & Premack, D. (1972). Teaching Language to an Ape. *Scientific American*, 227(4), 92–99.
Pulliainen, E. (1967). A Contribution to the Study of the Social Behavior of the Wolf. *American Zoologist*, 7, 313–317.
Pye, C. (1991). The Acquisition of K'iche' (Maya). In D. I. Slobin (Ed.), *The Crosslinguistic Study of Language Acquisition*, Vol. 3. (pp. 221–308). Hillsdale, New Jersey: Lawrence Erlbaum.
Pytte, C. L., Rusch, K. M., & Ficken, M. S. (2003). Regulation of vocal amplitude by the blue-throated hummingbird, Lampornis clemenciae. *Animal Behaviour*, 66, 703–710.
Pytte, C. L., Ficken, M. S., & Moiseff, A. (2004). Ultrasonic singing by the blue-throated hummingbird: a comparison between production and perception. *Journal of Comparative Physiology*, 8, 665–673.
Querleu, D., Lefebvre, C., Titran, M., Renard, X., Morillion, M., & Crepin, G. (1984). Reaction of the newborn infant less than 2 hours after birth to the maternal voice. *J. Gynecol. Obstet. Biol. Reprod*. 13, 125–134.
Radford, A. N. (2005). Group-specific vocal signatures and neighbour-stranger discrimination in the cooperatively breeding green woodhoopoe. *Animal Behaviour*, 70, 1227–1234.
Ramakrishnan, U., & Coss, R. G. (2000). Recognition of heterospecific alarm vocalization by Bonnet Macaques (Macaca radiata). *Journal of Comparative Psychology*, 114, 3–12.
Ramus, F., Hauser, M. D., Miller, C., Morris, D., & Mehler, J. (2000). Language Discrimination by Human Newborns and by Cotton-Top Tamarin Monkeys. *Science*, 288, 349–351.
Randall, J. A., & Matocq, M. D. (1997). Why do kangaroo rats (Dipodomys spectabilis) footdrum at snakes? *Behavioral Ecology*, 4, 404–413.
Range, F., Huber, L., & Heyes, C. (2011). Automatic imitation in dogs. *Proceedings of the Royal Society B*, 278, 211–217.
Regen, J. (1913). Über die Anlockung des Weibchens von Gryllus campestris L. durch telephonich übertragene Stridulationslaute des Männchens. *Arch. Physiol. Menchen und Tiere*, 155, 193–200.
Reiss, D., & McCowan, B. (1993). Spontaneous vocal mimicry and production by bottlenose dolphins (Tursiops truncatus): evidence for vocal learning. *Journal of Comparative Psychology*, 107, 301–312.
Reiss, D., McCowan, B., & Marino, L. (1997). Communicative and other cognitive characteristics of bottlenose dolphins. *Trends in Cognitive Sciences*, 1, 140–145.
Rekwot, P. I., Ogwu, D., Oyedipe, E. O., & Sekoni, V. O. (2001). The role of pheromones and biostimulation in animal reproduction. *Animal reproduction science*, 65, 157–170.
Remland, M. S., Jones, T. S., & Brinkman, H. (1995). Interpersonal distance, body orientation, and touch: effects of culture, gender, and age. *Journal of Social Psychology*, 135, 281–297.
Rendall, D., Owren, M. J., & Rodman, P. S. (1998). The role of vocal tract filtering in identity cueing in rhesus monkey (Macaca mulatta) vocalizations. *Journal of the Acoustic Society of America*, 103, 602–614.

Rendall, D., Rodman, P. S., & Emond R. E. (1996). Vocal recognition of individuals and kin in free-ranging rhesus monkeys. *Animal Behaviour*, 51, 1007–1015.

Rendell, L., & Whitehead, H. (2001). Culture in whales and dolphins. *Behavioural and Brain Sciences*, 24, 309–382.

Rendell, L., & Whitehead, H. (2004). Do sperm whales share coda vocalizations? Insights into coda usage from acoustic size measurement. *Animal Behaviour*, 67, 865–874.

Reynolds, V. (1963). An outline of the behaviour and social organization of forest-living chimpanzees. *Folia Primat*, 1, 95–102.

Reynolds, V., & Reynolds, F. (1965). Chimpanzees of the Budongo Forest. In I. DeVore, (Ed.), *Primate Behaviour: Field Studies of Monkeys and Apes*. (pp. 368–424). New York: Holt, Rinehart & Winston.

Riede, T., & Fitch, T. (1999). Vocal tract length and acoustics of vocalization in the domestic dog (*Canis familiaris*). *The Journal of Experimental Biology*, 202, 2857–2867.

Riede, T., Owren, M. J., & Arcadi, A. C. (2004). Nonlinear Acoustics in Pant Hoots of Common Chimpanzees (Pan troglodytes): Frequency jumps, subharmonies, biphonation, and deterministic chaos. *American Journal of Primatology*, 64, 277–291.

Riedel, J., Schumann, K., Kaminski, J., Call, J., & Tomasello, M. (2008). The early ontogeny of human-dog communication. *Animal Behaviour*, 75, 1003–1014.

Rizzolatti, G., & Arbib, M. A. (1998). Language within our grasp. *Trends in Neuroscience*, 21, 188–194.

Rizzolatti, G., Fadiga, L., Fogassi, L., & Gallese, V. (2002). From mirror neurons to imitation: Facts and speculations. In A. Meltzoff, & W. Prinz (Eds.), *The Imitative Mind. Development, evolution, and brain bases*. (pp. 247–266). Cambridge: Cambridge University Press.

Robinson, W. P. (1972). *Language and Social Behaviour*. Harmondsworth, UK: Penguin.

Robson, K. S. (1967). The role of Eye-to-Eye Contact in Maternal-Infant Attachment. *Journal of Child Psychology and Psychiatry*, 8, 13–25.

Rodseth, L., Wrangham, R. W., Harrigan, A. M., Smuts, B. B., Dare, R., Fox, R., King, B. J. Lee, P. C., Foley, R. A., Muller, J. C., Otterbein, K. F., Strier, K. B., Turke, P. W., & Wolpoff, M. H. (1991). The Human Community as a Primate Society. *Current Anthropology*, 32, 221–254.

Roeper, T. (2011). The acquistion of recursion: how formalism articulates the child's path *Biolinguistics*, 5, 1–2, 57–86.

Rooney, N. J., Bradshaw, J. S., & Robinson, I. H. (2001). Do dogs respond to play signals given by humans? *Animal Behaviour*, 61, 715–722.

Rossi, A. P., & Ades, C. (2008). A dog at the keyboard: using arbitrary signs to communicate requests. *Animal Cognition*, 11, 329–338.

Rothman, R. J., & Mech, L. D. (1979). Scent-marking in the lone wolves and newly formed pairs. *Animal Behaviour*, 27, 750–760.

Rotondo, J. L., & Boker, S. M. (2002). Behavioral synchronization in human conversational interaction. In M. I. Stamenov, & V. Gallese (Eds.), *Mirror Neurons and the Evolution of Brain and Language*. (pp. 151–162). Amsterdam: John Benjamins.

Roug, L., Landberg, I., & Lundberg, L-J. (1989). Phonetic development in early infancy: A study of four Swedish children during the first eighteen months of life. *Journal of Child Language*, 16, 19–40.

Rubenstein, D. I., & Hack, M. A. (1992). Horse signals: the sounds and scents of fury. *Evolutionary Ecology*, 6, 254–260.

Rumbaugh, D. M. (Ed.). (1977). *Language learning by a chimpanzee: The Lana project*. New York. Academic Press.

Sankar, R., & Archunan, G. (2004). Flehmen response in bull: role of vaginal mucus and other body fluids of bovine with special reference to estrus. *Behavioural Processes, 67*, 81–86.

Savage-Rumbaugh, E. S. (1986). *Ape language: From conditioned response to symbol.* New York: Columbia University Press.

Savage-Rumbaugh, E. S., & Lewin, R. (1994). *Kanzi. The ape at the brink of the human mind.* London: Doubleday.

Savage-Rumbaugh, E. S., McDonald, K., Sevcik, R. A., Hopkins, W. D., & Rupert, E. (1986). Spontaneous symbol acquisition and communicative use by pygmy chimpanzees (Pan paniscus). *Journal of Experimental Psychology: General, 115*, 211–235.

Savage-Rumbaugh, E. S., Sevcik, R. A., Rumbaugh, D. M., & Rubert, E. (1985). The capacity of animals to acquire language: do species differences have anything to say to us? *Phil. Trans. R. Soc. Lond. 308*, 177–185.

Savolainen, P. (2007). Domestication of dogs. In P. Jensen (Ed.), *The behavioural biology of dogs.* (pp. 21–37). Wallingford, Oxfordshire: CABI International Press.

Sayers, B. (1995). Wik-Mungkan. In N. Thieberger, & W. McGregor (Eds.), *Macquarie. Aboriginal words.* (pp. 352–373). Macquiarie University. Australia.

Sayigh, L. S., Tyack, P. L., Wells, R. S., & Scott, M. D. (1990). Signature whistles of free-ranging bottlenose dolphins *Tursiops truncatus*. Mother-offspring comparisons. *Behavioral Ecology and Sociobiology, 26*, 247–260

Sayigh, L. S., Tyack, P. L., Wells, R. S., Solow, A. R., Scott, M. D., & Irvine, A. B. (1998). Individual recognition in wild bottlenose dolphins: a field test using playback experiments. *Animal Behaviour, 57*, 41–50.

Schafer, G., & Plunkett, K. (1998). Rapid Word Learning by Fifteen-Months-Olds under Tightly Controlled Conditions. *Child Development, 69*, 309–320.

Scharff, C., & Haesler, S. (2005). An evolutionary perspective on FoxP2: strictly for the birds? *Current Opinion in Neurobiology, 15*, 694–703.

Schenkel, R. (1947). Ausdrucksstudien an Wölfen. *Behaviour, 1*, 81–129.

Schenkel, R. (1967). Submission: Its features and function in the wolf and dog. *American Zoologist, 7*, 319–329.

Scherer, K. R. (2003). Vocal communication of emotion: A review of research paradigms. *Speech Communication, 40*, 226–256.

Schevill, W. E., & Lawrence, B. (1949). Underwater listening to the white porpoise (Delphinapterus leuca). *Science, 109*, 143–144.

Schiffrin, D. (1974). Handwork as Ceremony: The Case of the Handshake. *Semiotica, 12*, 189–202.

Schino, G., Tiddi, B., & Polizzi Di Sorrentino, E. (2006). Simultaneous classification by rank and kinship in Japanese macaques. *Animal Behaviour, 71*, 1069–1074.

Schüch, W., & Barth, F. G. (1990). Vibratory communication in a spider: Female responses to synthetic male vibrations. *Journal of Comparative Physiology A, 166*, 817–826.

Schötz, S. (2006). Perception, Analysis and Synthesis of Speaker Age. Diss. Centre for Languages and Linguistics. Lund University.

Scott, J. P., & Fuller, J. L. (1965). *Genetics and the Social Behavior of the Dog.* Chicago: University of Chicago Press.

Scott, N. M., & Pika, S. (2012). A call for conformity. Gesture studies in human and non-human primates. In S. Pika, & K. Liebal (Eds.), *Developments in Primate Gesture Research.* (pp. 147–163). Amsterdam: John Benjamins.

Sèbe, F., Nowak, R., Poindron, P., & Aubin, T. (2007). Establishment of vocal communication and discrimination between ewes and lamb in the first two days after parturition. *Developmental Psychobiology*, 49, 375–386.

Sebeok. T. A. (1972). *Perspectives in zoosemiotics*. The Hague: Mouton.

Seeley, T. D. (1992). The tremble dance of the honeybee: message and meanings. *Behavioral Ecology and Sociobiology*, 31, 375–383.

Segerdahl, P., Fields, W., & Savage-Rumbaugh, E. S. (2005). *Kanzi's primal language: The Cultural Initiation of Primates into Language*. Hampshire; Palgrave Macmilian.

Serpell, J. (1995). *The domestic dog, its evolution, behaviour and interactions with people*. Cambridge: Cambridge University Press.

Seyfarth, R. M., & Cheney, D. L. (1986). Vocal development in vervet monkey grunts. *Animal Behaviour*, 34, 1640–1658.

Seyfarth, R. M., Cheney, D. L., & Marler, P. (1980a). Vervet monkey alarm calls: semantic communication in a free-ranging primate. *Animal Behaviour*, 28, 1070–1094.

Seyfarth, R. M., Cheney, D. L., & Marler, P. (1980b). Monkey responses to three different alarm calls: Evidence for predator classification and semantic communication. *Science*, 210, 801–803.

Shannon, C. E., & Weaver, W. (1949). *The mathematic theory of communication*. Urbana, IL.: University of Illinois Press.

Sibley, C. G., & Monroe, B. L. Jr. (1990). *Distribution and taxonomy of the birds of the world*. New Haven: Yale University Press.

Siebeck, U. E. (2004). Communication in coral reef fish: the role of ultraviolet colour patterns in damselfish territorial behaviour. *Animal Behaviour*, 68, 273–282.

Simpson, J. (2002). *A Learner's Guide to Warumungu*. Alice Springs. IAD Press.

Skinner, B. F. (1938). The behavior of organisms. New York: Appleton-Century-Crofts.

Slabbekoorn, H., & Peet, M. (2003). Birds sing at a higher pitch in urban noise. *Nature*, 424, 267.

Slabbekoorn, H., & den Boer-Visser, A. (2006). Cities change the songs of birds. *Current Biology*, 16, 2326–2331.

Slater, P. J. B., Ince, S. A., & Colgan, P. W. (1980). Chaffinch song types: their frequencies in the population and distribution between the repertoires of different individuals. *Behaviour*, 75, 207–218.

SLI Consortium. (2002). A Genomewide Scan Identifies Two Novel Loci Involved in Specific Language Impairment. *American Journal of Human Genetics*, 70, 384–398.

Slobin, D. I. (1996). From "thought and language" to "thinking for speaking" In J. J. Gumperz, & S. C. Levinson (Eds.), *Rethinking linguistic relativity*. (pp. 70–96). Cambridge: Cambridge University Press.

Slobin, D. I., & Aksu, A. A. (1982). Tense, aspect and modality in the use of the Turkish evidential. In P. J. Hopper (Ed.), *Tense-aspect: Between semantics & pragmatics*. (pp. 185–200). Amsterdam: John Benjamins.

Slobodchikoff, C. N., Kiriazis, J., Fisher, C., & Creef, E. (1991). Semantic information distinguishing individual predators in the alarm calls of Gunnison's prairie dog. *Animal Behaviour*, 42, 713–719.

Smith, J. L. D., McDougal, C., & Miquelle, D. (1989). Scent marking in free-ranging tigers, *Panthera tigris*. *Animal Behaviour*, 37, 1–10.

Smith, T. D., Bhatnagar, K. P., Shimp, K. L., Kinzinger, J. H., Bonar, C. J., Burrows, A. M., Mooney, M. P., & Siegel, M. I. (2002). Histological definition of the vomeronasal organ in humans and chimpanzees, with a comparison to other primates. *The anatomical record*, 267, 166–176.

Smolker, R., Mann, J., & Smuts, B. (1993). Use of signature whistles during separations and reunions between bottlenose dolphin mothers and infants. *Behavioral Ecology and Sociobiology*, 33, 393–402.

Smoski, M. J., & Bachorowski, J. A. (2003). Antiphonal laughter between friends and strangers. *Cognition and Emotion, 17*, 327–340.

Snowdon, C. T., & Pickhard, J. J. (1999). Family feuds: Severe Aggression among cooperatively breeding cotton-top tamarins. *International Journal of Primatology, 20*, 651–663.

Snowdon, C. T. & Elowson, A. M. (2001). 'Babbling' in Pygmy Marmosets: development after infancy. *Behaviour, 138*, 1235–1248.

Snowdon, C. T., & de la Torre, S. (2002). Multiple environmental contexts and communication in pygmy marmosets (*Cebuella pygmaea*). *Journal of Comparative Psychology, 116*, 182–188.

Soltis, J. Leong, K., & Savage, A. (2005a). African elephant vocal communication I: antiphonal calling behaviour among affiliated females. *Animal Behaviour, 70*, 579–587.

Soltis, J. Leong, K., & Savage, A. (2005b). African elephant vocal communication II: rumble variation reflects the individual identity and emotional state of callers. *Animal Behaviour, 70*, 589–599.

Sonesson, G. (2006). The meaning of meaning in biology and cognitive science: A semiotic reconstruction. *Sign Systems Studies, 34*(1), 135–213.

Soproni, K., Miklósi, Á., Tópál, J., & Csányi, V. (2001). Comprehension of Human Communicative Signs in Pet Dogs (Canis familiaris). *Journal of Comparative Psychology, 115*, 122–126.

Soproni, K., Miklósi, Á., Topíl, J., & Csányi, V. (2002). Dogs' (*Canis familiaris*) responsiveness to human pointing gestures. *Journal of Comparative Psychology, 116*, 27–34.

Spady, T. C., & Ostrander, E. A. (2008). Canine Behavioural Genetics: Pointing Out the Phenotypes and Herding up the Genes. *The American Journal of Human Genetics, 82*, 10–18.

Stamenov, M. I. (2002). Some features that make mirror neurons and human language faculty unique. In M. I. Stamenov, & V. Gallese (Eds.), *Mirror Neurons and the Evolution of Brain and Language.* (pp. 249–271). Amsterdam: John Benjamins.

Stamenov, M. I., & Gallese, V. (Eds.). (2002). *Mirror Neurons and the Evolution of Brain and Language.* Amsterdam: John Benjamins.

Steinhauer, K., White, E. J., & Drury, J. E. (2009). Temporal dynamics of late second language acquisition evidence from event-related brain potentials. *Second Language Research, 25*, 13–41.

Stevens, D. A. (1975). Laboratory methods for obtaining olfactory discrimination in rodents. In D. G. Moulton, A. Turk, & J. W. Johnston (Eds.), *Methods in Olfactory Research.* (pp. 375-394). New York: Academic Press.

Stevens, J. M. G., Vervaecke, H., DeVries, H., & Van Elsacker, L. (2006). Social structure in Pan paniscus: testing the female bonding hypothesis. *Primates, 47*, 210–217.

Stockhorst, U., & Pietrowsky, R. (2004). Olfactory perception, communication, and the nose-to-brain pathway. *Physiology & Behavior, 83*, 3–11.

Stopka, P., & Graciasová, R. (2001). Conditional allogrooming in the herb-field mouse. *Behavioural Ecology, 12*(5), 584–589.

Struhsaker, T. T. (1967). Auditory communication among vervet monkeys (*Cercopithecus aethiops*). In S. A. Altmann (Ed.), *Social Communication among Primates.* (pp. 281–324). Chicago: University of Chicago Press.

Strum, S. C., & Fedigan, L. M. (2000). Changing views of primate society: a situated North American view. In S. C. Strum, & L. M. Fedigan (Eds.), *Primate Encounters: Models of Science, Gender, and Society.* (pp. 3–49). Chicago: University of Chicago Press.

Sturdy, C. B., & Mooney, R. (2000). Bird communication: Two voices are better than one. *Current Biology, 10*, R6343–R636.

Studdert-Kennedy, M. (1998). The particulate origin of language generativity: from syllable to gesture. In J. R. Hurford, M. Studdert-Kennedy, & C. Knight (Eds.), *Approaches to the Evolution of Language* (pp. 202-221). Cambridge: Cambridge University Press.

Suguira, H. (1998). Matching of acoustic features during the vocal exchange of coo calls by Japanese macaques. *Animal Behaviour*, 55, 673-687.

Suthers, R. A. (1990). Contributions to birdsong from the left and right sides of the intact syrinx. *Nature*, 347, 473-477.

Suzuki, A. (1969). An ecological study of chimpanzees in a savannah woodland. *Primates*, 10, 103-148.

Sweeney, A., Jiggins, C., & Johnsen, S. (2003). Polarized light as a butterfly-mating signal. *Nature*, 423, 31.

Szetei, V., Miklósi, A., Topál, J., & Csányi, V. (2003). When dogs seem to lose their nose: an investigation on the use of visual and olfactory cues in communicative context between dog and owner. *Applied Animal Behaviour Science*, 83, 141-152.

Szmrecsanyi, B. (2005). Language users as creatures of habit: A corpus-based analysis of persistence in spoken English. *Corpus Linguisics and Linguistic Theory*, 1, 113-150.

Takahashi, M., Arita, H., Hiraiwa-Hasegawa, M., & Hasegawa, T. (2008). Peahens do not prefer peacocks with more elaborate trains. *Animal Behaviour*, 75, 1209-1219.

Tallerman, M. (Ed.). (2005). *Language Origins: Perspectives on Evolution*. Oxford: Oxford University Press.

Tardif, T., Liang, W., Zhang, Z., Fletcher, P., Kaciroti, N., & Marchman, V. A. (2008). Baby's first 10 words. *Developmental Psychology*, 44, 929-938.

Tavolga, W. N. (Ed.). (1964). *Marine Bio-acoustics*. New York: Pergamon Press.

Taylor, A. M., Reby, D. & McComb, K. (2010). Size communication in domestic dog, *Canis familiaris*, growls. *Animal Behaviour*, 79, 205-210.

Terrace, H. S. (1979). *Nim: A Chimpanzee Who Learned Sign Language*. New York. Knopf.

Terrace, H. S., Petitto, L. A., Sanders, R. J., & Beaver, T. G. (1979). Can an ape create a sentence? *Science*, 206, 891-206.

Terrazas, A., Nowak, R., Serafin, N., Ferreira, G., Lévy, F., & Poindron, P. (2002). Twenty-Four-Hour-Old Lambs Rely more on Maternal Behavior Than on the Learning of Individual Characteristics to Discriminate Between Their Own and an Alien Mother. *Developmental Psychobiology*, 40, 408–418.

Thomsen, F., Franck, D., & Ford, J. K. B. (2002). On the communicative significance of whistles in wild killer whales (*Orcinus orca*). *Naturwissenschaften*, 89, 404-407.

Thorpe, W. H. (1951). The learning abilities of birds. *Ibis*, 93, 1-52, 252-296.

Thorpe, W. H. (1958). The learning of song patterns by birds, with special reference to the song of the chaffinch *Fringilla coelebs*. *Ibis*, 100, 535-570.

Thorpe, W. H. (1972). The Comparison of Vocal Communication in Animals and Man. In R. A. Hinde (Ed.), *Non-Verbal Communication*. (pp. 27-47). Cambridge: Cambridge University Press.

Tinbergen, N. (1951). *The Study of Instinct*. Oxford: Claredon Press.

Tinbergen, N. (1960). Comparative studies of the behaviour of gulls (Laridae): A progress report. *Behaviour*, 15, 1-70.

Tirindelli, R., Mucignat-Caretta, C., & Ryba, N. J. B. (1998). Molecular aspects of pheromonal communication via the vomeranasal organ of mammals. *Trends Neurosci.*, 21, 482-486.

Todt, D. (2004). From birdsong to speech: a plea for comparative approaches. *Annals of the Brazilian Academy of Sciences*, 76.

Todt, D., & Hultsch, H. (1996). Acquisition and performance of repertoires: ways of coping with diversity and versatility. In D. E. Kroodsma & H. Miller (Eds.), *Ecology and evolution of communication in birds.* (pp. 79–96). Ithaca, New York: Cornell University Press.

Todt, D., & Hultsch, H. (1998). How songbirds deal with large amounts of serial information: retrieval rules suggest a hierarchical song memory. *Biological Cybernetics, 79,* 487–500.

Tomasello, M. (1992). *First Verbs: A Case Study in Early Grammatical Development.* Cambridge: Cambridge University Press.

Tomasello, M. (2003). *Constructing a language. A Usage-Based Theory of Language Acquisition.* Cambridge, Mass.: Harvard University Press

Tomasello, M. (2008). *Origins of Human Communication.* Cambridge, Mass.: MIT Press.

Tomasello, M., Call, J., & Hare, B. (1998). Five primate species follow the visual gaze of conspecifics. *Animal Behaviour, 55,* 1063–1069.

Tomasello, M., & Camaioni, L. (1997). A comparison of the gestural communication of apes and human infants. *Human Development, 40,* 7–24.

Tomasello, M., & Moll, H. (2010). The Gap is social: Human shared Intentionality and Culture. In P. M. Kappeler, & J. B. Silk (Eds.), *Mind the gap: Tracing the Origins of Human Universals.* (pp. 331–349). Berlin. Heidelberg: Springer-Verlag.

Tomonaga, M. (2006). Development of Chimpanzee Social Cognition in First 2 Years. In T. Matsuzawa, M. Tomonaga, & M. Tanaka (Eds.), *Cognitive Development in Chimpanzees.* (pp. 182–197). Tokyo: Springer Verlag.

Tomonaga, M., Tanaka, M., Matsuzawa, T., Myowa-Yamakoshi, M., Kosugi, D., Mizuno, Y., Okamoto, S., Yamaguchi, M., & Bard, K. (2004). Development of social cognition in infant chimpanzees (Pan troglodytes): Face recognition, smiling, gaze, and the lack of triadic interactions. *Japanese Psychological Research, 46,* 227–235.

Torriani, M. V. G., Vannoni, E., & McElligott, A. G. (2006). Natural History Miscellany. Mother-Young Recognition in an Ungulate Hider Species: A Unidirectional Process. *The American Naturalist, 168,* 412–420.

Trevarthen, C. (1977). Descriptive Analyses of Infant Communicative Behaviour. In H. R. Schaffer (Ed.), *Studies in the Mother-Infant Interaction.* (pp. 227–270). London: Academic Press.

Trevarthen, C. (1979). Communication and cooperation in early infancy. In M. Bullowa (Ed.), *Before Speech: The Beginnings of Human Communication.* (pp. 321–347). Cambridge: Cambridge University Press.

Triefenback, F., & Zakon, H. (2003). Effects of sex, sensitivity and status on cue recognition in the weakly electric fish *Apteronotus leptorhynchus. Animal Behaviour, 65,* 19–28.

Tschanz, B. B. (1959). Zur Brutbiologie der Trottellumme (*Uria aalge aalge* Pont.). *Behaviour, 14,* 1–100.

Tsushima, T., Takizawa, O., Sasaki, M., Siraki, S., Nishi, K., Kohno, M., Menyuk, P., & Best, C. (1994). Discrimination of English /r-l/ and /w-y/ by Japanese infants at 6–12 months: Language specific developmental changes in speech perception abilities. Paper presented at *International Conference on Spoken Language Processing, 4.* Yokohama. Japan.

Tyack, P. (1983). Differential response of humpback whales to playback of songs or social sounds. *Behavi Ecol Sociobiol, 8,* 105–116.

Tyack, P. (1986). Population biology, social behaviour and communication in whales and dolphins. *TREE, 1,* 144–150.

Tyack, P. L. (1998). Acoustic communication under the sea. In S. L. Hopp, M. J. Owren, & C. S. Evans (Eds.), *Animal Acoustic Communication. Sound Analysis and Research Methods.* (pp. 163–220). Berlin: Springer-Verlag.

van Baaren, R. B., Holland, R. W., Steenaert, B., & van Knippenberg, A. (2003). Mimicry for money: Behavioral consequences of imitation. *Journal of Experimental Social Psychology*, 39, 393–398.

Veá, J.J., & Sabater-Pi, J. (1998). Spontaneous pointing behaviour in the wild pygmy chimpanzee (*Pan paniscus*). *Folia primatologica*, 69, 289–290.

Ventureyra, V., Pallier, C., & Yoo, H.-Y. (2004). The loss of first language phonetic perception in adopted Koreans. *Journal of Neurolinguistics*, 17, 79–91.

Vergne, A. L., Avril, A., Martin, S., & Mathevon, N. (2007). Parent-offspring communication in the Nile crocodile *Crocodylus niloticus*: do newborns' calls show an individual signature? *Naturwissenschaften*, 94, 49–54.

Vernes, S. C., Nicod, J., Elahi, F. M., Coventry, J. A., Kenny, N., Coupe, A-M., Bird, L. E., Davies, K. E., & Fisher, S. E. (2006). Functional genetic analysis of mutations implicated in a human speech and language disorder. *Human Molecular Genetics*, 15(21), 3154–3167.

Vettin, J., & Todt, D. (2004). Laughter in conversation: features of occurrence and acoustic structure. *Journal of Nonverbal Behaviour*, 28, 93–115.

Vick, S-J., Waller, B. M., Parr, L. A., Smith Pasqualini, M. C., & Bard, K. A. (2007). Cross-species Comparison of Facial Morphology and Movement in Humans and Chimpanzees Using the Facial Action Coding System (FACS). *Journal of Nonverbal Behaviour*, 31, 1–20.

Vihman, M. M., Ferguson, C., & Elbert, M. (1986). Phonological development from babble to speech: common tendencies and individual differences. *Applied Psycholinguistics*, 7, 3–40.

Virányi, Z., Topál, J., Gácsi, M., Miklósi, Á., & Csányi, V. (2004). Dogs respond appropriately to cues of humans' attentional focus. *Behavioural Processes*, 66, 161–172.

Virányi, Z., Gácsi, M., Kubinyi, E., Topál, J., Belényi, B., Ujfalussy, D., & Miklósi, A. (2008). Comprehension of human pointing gestures in young human-reared wolves (*Canis lupus*) and dogs (*Canis familiaris*). *Animal Cognition*, 11, 373–387.

van Hooff, J. A. R. A. M. (1972). A comparative approach to the phylogeny of laughter and smile. In R. A Hinde (Ed.), *Non-verbal communication*. (pp. 209–238). Cambridge: Cambridge University Press.

Vormbrock, J. K., & Grossberg, J. M. (1988). Cardiovascular effects of human-pet dog interactions. *Journal of Behavioural Medicine*, 11, 509–517.

Wada, K., Sakaguchi, H., Jarvis, E. D., & Hagiwara, M. (2004). Differential Expression of Glutamate Receptors in Avian Neural Pathways for Learned Vocalization. *The Journal of Comparative Neurology*, 476, 44–64.

Waller, B. M., Vick, S.-J., Parr, L. A., Smith Pasqualini, M. C., Bard, K. A., & Gothard, K. (2006). Intramuscular electrical stimulation of facial muscles in humans and chimpanzees: Duchenne revisited and extended. *Emotion*, 6, 367–382.

Wayne, R. K., & Vilà, C. (2003). Molecular genetic studies of wolves. In L. D. Mech, & L. Boitani (Eds.), *Wolves. Behaviour, Ecology, and Conservation*. (pp. 218–238). Chicago: The University of Chicago Press.

Weale, E. (1997). From Babel to Brussels. Conference interpreting and the art of the impossible. In F. Poyatos (Ed.), *Nonverbal communication and translation: New perspectives and challenges in literature, interpretation and the media*. (pp. 295–312). Amsterdam: John Benjamins.

Weary, D. M. (1989). Categorical Perception of Bird Song: How do great tits (*Parus major*) Perceive Temporal Variation in Their Song?. *Journal of Comparative Psychology*, 103, 300–323.

Wedekind, C., Seebeck, T., Bettens, F., & Paepke, A. J. (1995). MHC-dependent mate preferences in humans. *Proceedings of R Soc Lond B*, 260, 245–249.

Werker, J. F. (1995). Exploring developmental changes in cross-language speech perception. In L. Gleitman (Ed.), *An invitation to Cognitive Science: Language.* (pp. 87–106). MIT Press.

Wexler, K. (2003). Lenneberg's Dream: Learning, Normal Language Development and Specific Language Development. In J. Schaffer, & Y. Levy (Eds.), *Language Competence Across Populations: Towards a Definition of Specific Language Impairment.* (pp. 11–61). London: Lawrence Erlbaum Associates.

Whaling, C. S., Solis, M. M., Doupe, A. J., Soha, J. A., & Marler, P. (1997). Acoussic and neural bases for innate recognition of song. *Proc. Natl. Acad. Sci. USA,* 94, 12694-12698.

White, S. A. (2001). Learning to communicate. *Current Opinion in Neurobiology,* 11, 510–520.

Whiten, A., Goodall, J., McGrew, W. C., Nishida, T., Reynolds, V., Sugiyama, Y., Tutin, C. E. G., Wrangham, R. W., & Boesch, C. (1999). Cultures in chimpanzees. *Nature,* 399, 682–685.

Whiten, A., Goodall, J., McGrew, W. C., Nishida, T., Reynolds, V., Sugiyama, Y., Tutin, C. E. G, Wrangham, R. W., & Boesch, C. (2001). Charting cultural variation in chimpanzees. *Behaviour,* 138, 1489–1525.

Wilcox, R. S. (1979). Sex discrimination in Gerris remigis: role of a surface wave signal. *Science,* 206, 1325–1327.

Wilson, D. R., Bayly, K. L., Nelson, X. J., Gillings, M., & Evans, C. S. (2008). Alarm calling best predicts mating and reproductive success in ornamented male fowl, *Gallus gallus. Animal Behaviour,* 76, 543–554.

Wimmer, H., & Perner, J. (1983). Beliefs about beliefs: representation and constraining function of wrong beliefs in young children's understanding of deception. *Cognition,* 13, 103–128.

Wobber, V., Hare, B., Koler-Matznick, J., Wrangham, R., & Tomasello, M. (2009). Breed differences in domestic dogs' (Canis familiaris) comprehension of human communicative signals. *Interaction Studies,* 10, 206–224.

Wray, A. (Ed.). (2002). *The transition to language.* Oxford: Oxford University Press.

Wyatt, T. D. (2003). *Pheromones and Animal Behaviour. Communication by Smell and Taste.* Cambridge: Cambridge University Press.

Yamashita, C. (1987). Field Observations and Comments on the Indigo Macaw (Anodorhynchus leari), a Highly Endangered Species from Northeastern Brazil. *Wilson Bulletin,* 99, 280–282.

Yeon, S. C. (2007). The vocal communication of canines. *Journal of Veterinary Behavior,* 2, 141–144.

Yin, S., & McCowan, B. (2004). Barking in domestic dogs: context specificity and individual identification. *Animal Behaviour,* 68, 343–355.

Yip, M. (2006). The search for phonology in other species. *Trends in Cognitive Sciences,* 10, 442–446.

Yurk, H., Barrett-Lennard, L., Ford, J. K. B., & Matkin, C. O. (2002). Cultural transmission within maternal lineages: vocal clans in resident killer whales in southern Alaska. *Animal Behaviour,* 63, 1103–1119.

Zeesman, S., Nowaczyk, M. J. M., Teshima, I., Roberts, W., Cardy, J. O., Brian, J., Senman, L., Feuk, L., Osborne, L. R., & Scherer, S. W. (2006). Speech and Language Impairment and Oromotor Dysphraxia Due to Deletion of 7q31 That Involves FOXPs. *American Journal of Medical Genetics,* 140A, 509–514.

Zetterholm, E. (2003). *Voice Imitation. A Phonetic Study of Perceptual Illusions and Acoustic Success.* Travaux de l'Institute de linguistique de Lund 44. Lund: Studentlitteratur.

Zimen, E. (1971). *Wölfe und Königspudel. Vergleichende Verhaltens-beobachtungen.* München: Piper & Co.

Zimen, E. (1981). *The Wolf. His place in the natural world.* London: Souvenir Press Ltd.

Zou, Z., & Buck, L. B. (2006). Combinatorial Effects of Odorant Mixes in Olfactory Cortex. *Science, 311*, 1477–1480.

Zuberbühler, K. (2000a). Referential labelling in Diana monkeys. *Animal Behaviour, 59*, 917–927.

Zuberbühler, K. (2000b). Interspecies semantic communication in two forest primates. *Proc. R. Soc. Lond. B, 267*, 713–718.

Zuberbühler, K. (2002). A syntactic rule in forest monkey communication. *Animal Behaviour, 63*, 293–299.

Zuberbühler, K., Noë, R., & Seyfarth, R. M. (1997). Diana monkey long-distance calls: messages for conspecifics and predators. *Animal Behaviour, 53*, 589–604.

Index

A
accommodation 48, 80, 158, 180, 188, 201
alignment 5, 21, 48, 55, 67, 74, 80, 168–169, 188–189, 201
allogrooming 35, 192, 201
antiphonal 27, 84, 169, 188
audience 8, 93, 103, 116–117, 136, 148–149, 170, 201
    audience design 8, 136

B
babbling 31, 57–59, 76, 100, 121, 129, 136, 178, 196–197
Baby Talk 21, 55, 72, 85, 159–160
bilingual 49–51, 135, 180, 196
Brown, Roger 52, 66, 130, 197

C
calls 4, 12–13, 20, 26–28, 31–32, 114–121, 136–137, 163, 165, 169–170, 180, 186, 188, 190, 201
    alarm calls 4, 13, 20, 26, 115–117, 120–121, 136–137, 188, 190
    food calls 12–13, 26, 31, 116–117, 120, 136–137, 170, 188
Campbell's monkey 110, 119, 122, 189
categorical perception 17, 56, 138, 194, 197, 201
chaffinch 7, 15, 168, 181
Cheney, Dorothy 13, 115–116, 136, 138
Chomsky, Noam 5, 15–16, 18, 77, 137
chorus 145–146, 151, 166, 189, 201
    chorus howling 145–146, 151
communicative functions 1–2, 53–54, 115–118, 131, 136–137, 170, 187, 190, 202
    see also language functions

cooperation 42, 70, 77, 79, 110, 117–118, 144, 147, 188, 198
cooperative breeding 19–20, 22, 112, 119, 121, 136, 138–139, 144, 165–166, 191, 195
counter-singing 168–169, 186
cricket 24, 26, 28
critical period 14, 16, 75, 134, 188, 201
crocodile 20, 27
culture 3, 21, 32, 55, 59, 61, 63, 72, 80–82, 86–88, 90–96, 98–100, 103, 124, 191

D
Darwin, Charles 4, 6–7, 18, 52, 80, 90, 92–94, 112, 182, 188, 191, 202
deictic 99, 101–102, 107, 201
dialect 3, 7, 15, 32, 48, 85, 120–121, 128, 152, 176, 180–182, 188, 200
Diana monkey 110, 116, 122
displacement 10–12, 100, 189–190
dolphin 3, 23, 25–31
domestic chicken 13, 25–26, 164, 169–170, 189–190, 203
duet 118–119, 121, 168–169, 173, 186, 189, 201

E
eavesdropping 8, 201
Ekman, Paul 80–81, 91, 97–100, 166, 201–203
elephant 25, 27–28, 35, 38–39, 41–42, 146, 188
embedded clause 14, 69, 185, 203
emotion 6, 13, 53, 81, 83–86, 90–94, 97, 99, 103–105, 107, 115, 120, 122–123, 136, 146, 150, 157–159, 198

F
fish 29–30, 34, 36, 38, 42, 94
fission-fusion 3, 30–31, 110–111, 137, 165, 188, 202
FOXP2 19, 71, 180
frequency code 55, 85, 90, 125, 150, 157–158, 194, 202
Friesen, Wallace 80–81, 91, 97–100, 201–203
Frisch, Karl v. 7, 23, 29, 42–44, 190

G
gaze 21, 79, 82, 87–88, 92–94, 97, 99, 107, 114, 117, 123–126, 128, 136, 148–149, 156, 158–159, 188, 191, 196, 203
    gaze following 21, 117, 123, 125–126, 188, 191
    mutual gaze 21, 93, 117, 125, 196
gibbon 109–110, 118–119, 121, 189, 203
glands 39–40, 97, 155, 183, 202
Goffman, Erving 89, 93, 108
Goodall, Jane 110–115, 120, 123, 128, 135, 137
greetings 3, 34–35, 53, 61, 81, 86, 89–91, 95, 111, 137, 150
grooming 18, 35, 86, 100, 109, 111, 119, 122–123, 125–128, 136–138, 192, 201

H
Hockett, Charles 10–11, 189–190, 199
honeybee 23, 25, 42–44
Hrdy, Sarah, Blaffer 20, 22, 112, 166, 195
hunting 139–140, 143–144, 147, 161, 165–166, 191–192, 195

I

imprinting 7, 14–15, 202

J

Jakobson, Roman 6, 12–13, 53

K

kinship 63–64, 77, 120
  kinship terminology 63
Kuhl, Patricia 17, 55–57, 160, 194, 197

L

language families 4, 49
language functions 6, 53, 202
larynx 57–58, 172, 193–194, 197, 202
laughter 84, 97, 103, 114, 188, 202
learning 11, 15–16, 19, 21–22, 30–32, 48, 55–62, 66, 72–76, 91, 106, 120–121, 123, 125, 127–129, 131, 133–134, 143–144, 148, 151, 158–161, 163–167, 171, 176–180, 183–186, 188, 191, 195–196, 198–199
lexigram 123–124, 133–134, 148–149
Lorenz, Konrad 7, 14–15, 18, 92, 140

M

Marler, Peter 13, 15, 117, 165–167, 169–170, 177, 179, 186, 189–190
marmoset 110, 112, 119, 121–122, 135, 196–197, 203
metaphors 104–105, 187
mirror neurons 21, 83, 159, 189, 202
mirroring 59, 81, 83–84

N

nonverbal communication 2–5, 34, 79–82, 89, 97–98, 100–101, 103, 107, 118
noun 67, 69, 105

O

onomatopoetic 26, 105–106, 114, 166
ontogeny 7–8, 177, 194, 200, 203

P

pant hoot 116, 119–120
parent-offspring communication 19–22, 196, 203
pheromones 38–41, 43–44, 46, 96, 202–203
phrase 17, 32, 63, 69, 73–74, 80, 118, 129, 174–176
playback 13, 24–25, 45, 113–116, 118, 120, 146, 151–152, 166–169, 171–172, 186, 188, 203
pointing 21, 30, 59, 76, 82, 86–88, 92, 99, 107, 117, 123–125, 129, 136, 141, 145, 147–149, 158, 160, 188
Processability Theory 73–74, 176
pronoun 55, 63, 67, 76, 160

R

recursion 14, 68–70, 185, 203
referential function 13, 43, 110, 115–118, 137–138, 147–149, 169–170, 188, 203

S

second language acquisition 16, 57, 70, 72–75, 77, 196
Seyfarth, Robert 13, 114–116, 136, 138
SLI (Specific Language Impairment) 70–71, 77, 107, 180, 203
smiling 3, 54, 81, 90–91, 99, 102
social function 6, 13, 35, 42, 53–55, 81–82, 84, 107, 110, 115, 118–119, 136–137, 146, 160, 163
song 15, 17, 19, 31–32, 75, 118–119, 121, 163–165, 167–169, 171–182, 186, 190, 195–196
  song learning 15, 19, 24, 75, 164, 176–167, 180, 196
  subsong 177–178, 196, 201
songbirds 15–16, 19, 24, 31, 71, 77, 163–165, 171–172, 174, 176–178, 180–181, 185, 190, 195–196
syllables 99, 151, 172, 174–176, 195
synchronization 4–5, 21, 48, 79, 83, 91, 99, 102, 107, 110, 118–119, 121, 169, 188–189, 199

syntax 18, 32, 70, 122, 130, 134, 138, 158, 174–176, 185, 188, 198, 204
syrinx 172, 203

T

tamarin 38, 110, 112, 194, 197, 203
theory of mind 69–70, 204
threat 34, 40, 55, 85, 90, 92–93, 104, 121, 125, 150, 154, 158, 194
Tinbergen, Nikolaas 7–8, 18, 25, 45, 100

U

Universal Grammar 5, 16, 18, 77, 204

V

verb 26, 54–55, 62, 67–68, 73, 104–105, 131
vervet monkeys 13, 25, 59, 110, 114, 116–117, 135, 138, 146, 190
vocabulary 12, 14, 59, 67, 72, 76, 114, 129–131, 134, 137, 159–160, 184
vocal tract 58, 85, 129, 151, 194

W

waggle dance 7, 43–44
whales 3, 23, 25–27, 29, 31–32
whistle 26, 28, 30–32, 157, 179, 183
words 4–6, 11–12, 14–15, 17, 21, 48, 52–56, 58–68, 72–74, 76–77, 79–80, 84, 86–87, 89, 93, 98, 100, 102–107, 129–131, 134–135, 138, 148–149, 158, 160, 175, 190, 196–197